Field Manual FM 3-11

Chemical, Biological, Radiological, and N

May 2019

United States Government
US Army

Foreword

As the Chemical Corps enters its second century of service in the United States Army, it must adapt to new threats and overcome 15 years of atrophy of chemical, biological, radiological, and nuclear (CBRN) skills within the Army that were attributed to operations in counterinsurgency. Throughout that time, the Chemical Corps approached the CBRN problem as it had for the past 100 years—avoid, protect, and decontaminate. These three insular capability areas provided a linear approach to detect existing hazards and offered a bypass of CBRN obstacles to enable the safe passage of forces. The approach to the CBRN program remained resistant to change, yet it ignored new forms of battle being exercised and conducted by major regional players (Russia, China, Iran, North Korea). The proliferation of new technologies, to include weapons of mass destruction (WMD) capabilities and materials, will remain a constant during the next 100 years. The Chemical Corps, in conjunction with the Army and joint force, must adapt and prepare for CBRN usage throughout the range of military operations.

The Chemical Corps exists to enable movement and maneuver to conduct large-scale ground combat operations in a CBRN environment. Friendly forces must retain freedom of action and be capable of employing the full breadth of capabilities within complex battlefield conditions, including CBRN environments. The latest addition of FM 3-0 and the Army concept of multidomain operations outline the essential elements for success against peer and near-peer adversaries. FM 3-11 is the answer to FM 3-0's requirement to adapt and to overcome in the plausibility of enemy CBRN usage to create conditions of overmatch. To support multidomain operations, the Army CBRN Regiment must be agile and adaptive, capable of employing its full capabilities to enable the Army to win in a complex CBRN environment. It is essential that U.S. forces build and maintain operational readiness whilst restoring CBRN mastery at tactical and operational levels of war. CBRN must be integrated into all training facets as a condition on the battlefield that the enemy will leverage to establish and maintain a position of relative advantage.

FM 3-11 provides a thorough doctrinal approach for the tactical and operational levels of war to provide tailorable, scalable CBRN capabilities across operations. It acknowledges that CBRN units must be integrated at multiple tactical and operational headquarters to enhance their capabilities to counter WMD and retain operational flexibility. Most importantly, it transforms the CBRN enterprise to being offensive and proactive, interdicting CBRN before employment rather than simply practicing avoidance.

The time has come to make CBRN integration into training a priority for every leader. Building agile and adaptive leaders who can win means forcing them to adapt to the complex rigors of a CBRN environment that provides a substantial advantage to threat forces. FM 3-11 provides leaders and staffs the ability to incorporate CBRN units into their formations and to defeat the enemies' ability to use CBRN hazards effectively in large-scale ground combat.

Antonio Munera

ANTONIO V. MUNERA
BRIGADIER GENERAL
CBRN COMMANDANT

Field Manual
No. 3-11

Headquarters
Department of the Army
Washington, DC, 23 May 2019

Chemical, Biological, Radiological, and Nuclear Operations

Contents

DISTRIBUTION RESTRICTION: Approved for public release; distribution is unlimited.

*This publication supersedes FM 3-11, dated 1 July 2011.

Figures

Tables

Preface

FM 3-11 provides commanders and staffs with overarching chemical doctrine for operations to assess, protect, and mitigate the entire range of CBRN threats and hazards—including support to countering weapons of mass destruction (CWMD) activities in all operational environments. It addresses principles, fundamentals, planning, operational considerations, and training and support functions. It provides a common framework and language for CBRN operations and constitutes the doctrinal foundation for developing other fundamentals and tactics, techniques, and procedures detailed in subordinate doctrine manuals. This manual is a key integrating publication that links the doctrine for the CBRN units and staffs with Army operational doctrine and joint doctrine.

The principal audience for FM 3-11 is commanders, staffs, and leaders of theater armies, corps, divisions, and brigades as well as CBRN units that integrate capability into those formations. However, FM 3-11 is applicable to all members of the profession of arms. To comprehend the doctrine in FM 3-11 readers must first understand the fundamentals of unified land operations described in ADP 3-0 and in FM 3-0. The reader must also understand the language of tactics and the fundamentals of the offense and defense described in ADP 3-90, and be familiar with operational terms and graphics described in ADP 1-02. Commanders and staffs of Army headquarters should also refer to applicable joint or multinational doctrine concerning the range of military operations (ROMO) and joint or multinational forces.

Commanders, staffs, and subordinates ensure that their decisions and actions comply with applicable United States, international, and in some cases host-nation laws and regulations. Commanders at all levels ensure that their Soldiers operate in accordance with the law of war and the rules of engagement. (See FM 27-10.)

FM 3-11 uses joint terms where applicable. Selected joint and Army terms and definitions appear in both the glossary and the text. Terms for which FM 3-11 is the proponent publication (the authority) are italicized in the text and are marked with an asterisk (*) in the glossary. Terms and definitions for which FM 3-11 is the proponent publication are boldfaced in the text. For other definitions shown in the text, the term is italicized and the number of the proponent publication follows the definition.

FM 3-11 applies to the Active Army, Army National Guard/Army National Guard of the United States and United States Army Reserve unless otherwise stated.

The proponent of FM 3-11 is the United States Army Chemical, Biological, Radiological, and Nuclear School (USACBRNS). The preparing agency is the Maneuver Support Center of Excellence (MSCoE) Capabilities Development and Integration Directorate (CDID); Concepts, Organizations, and Doctrine Development Division (CODDD); Doctrine Branch. Send comments and recommendations on DA Form 2028 (*Recommended Changes to Publications and Blank Forms*) to Commandant, USACBRNS, ATTN: ATZT-CDC, 14000 MSCoE Loop, Suite 235, Fort Leonard Wood, MO 65473-8929; by e-mail to usarmy.leonardwood.mscoe.mbx.cdidcodddcbrndoc@mail.mil; or submit an electronic DA Form 2028.

Introduction

Throughout the Cold War and into the 1990s, United States (U.S.) training and doctrine acknowledged that enemy CBRN usage to shape the battlefield was an imminent reality. Training at the Combined Arms Training Centers frequently involved reacting to chemical strikes within division and brigade support areas, while maintaining operational reach in large-scale ground combat. Infantry and armor formations pushed the limit of their capabilities against robust CBRN strikes within decisive action.

Over the past 15 years of conflict in the Middle East, U.S. defense policy reflected a reduction in concern for CBRN use by contemporary threats. With operational and strategic emphasis on counterinsurgency operations in the Middle East, this belief was affirmed by the asymmetric threat's lack of technical capability to effectively manufacture and deliver CBRN, short of already-produced agents and hazardous materials. The result was an aggregate regression of CBRN training and readiness across U.S. Army formations.

Today, peer threats recognize the U.S. forces reduced capability to operate in a CBRN environment. With an integrated air defense capability as well as superiority in fires, adversaries might leverage multiple complex mission variables, such as CBRN, to deny key terrain, isolate friendly forces, and induce battlefield complexity, creating conditions for regional overmatch.

In 2015, Russia realigned its CBRN forces, with a CBRN brigade structured to support every combined arms and tank army. During annual training exercises the Russian Army trains extensively in CBRN conditions. Much like the Soviet Army of the past, they see CBRN as a condition on the battlefield that when employed on constrictive, canalizing terrain, is exploitable, and provides conditions for force overmatch. North Korea also maintains a robust CBRN program that threatens the Republic of Korea and surrounding countries, focusing their recent efforts on theater and intercontinental ballistic missiles that plausibly could strike the U.S. mainland. Iran remains a destabilizing regional adversary in the Middle East, with nuclear ambitions and a government adversarial to U.S. interests. China maintains a covert CBRN program that continues to grow and may be used in future conflicts.

Each of these CBRN capable near-peer threats represents a challenge to the U.S. Army's readiness across the range of military operations. It is highly anticipated that these threats will use multiple approaches and in different phases to leverage CBRN to their advantage.

The Chemical Corps exists to provide the freedom of action for Army units operating in CBRN environments. The U.S. Army's CBRN force is an agile, adaptive team that provides critical capability to enable the Army to fight and win in a complex CBRN environment. This manual describes how to employ CBRN capabilities to enable freedom of action within U.S. Army formations and how CBRN staffs can provide expertise to maximize survivability in order to prevail in large-scale combat operations and exploit enemy use of CBRN.

To understand FM 3-11, the reader must first understand the doctrinal fundamentals contained in ADPs 1-02, 3-0, 3-90, 5-0, and 6-0; and FMs 3-0 and 3-90-1. The reader should know the activities described in FM 3-0 in order to understand how CBRN operations support the elements of decisive action and their subordinate tactical, enabling, and sustaining tasks. FM 3-11 is meant to be the CBRN companion publication to FM 3-0.

FM 3-11 serves as a foundation of knowledge and provides the professional language that guides how CBRN Soldiers perform tasks related to the Army's role—the employment of land power to support joint operations.

This revision is based on the new focus of the Army toward large-scale combat operations. It represents a significant change from the previous version. The most significant changes to FM 3-11—

- Establishes and describes the CBRN core functions of assess, protect, and mitigate and discontinues use of the eight military mission areas.
- Provides basic structure for CBRN's role in the Army's operational concept of unified land operations.

- Describes the organizations, capabilities and training of CBRN operations supporting the tasks of decisive action.
- Provides an understanding of how CBRN forces directly support maneuver forces in large-scale combat operations.
- Emphasizes the importance of preparation and training for large-scale combat operations in CBRN environments.

The following is a brief introduction and summary of the chapters.

- **Chapter 1** provides a framework for the core concepts of the FM and the CBRN Corps. It discusses core functions, operational framework, and the operational environment (OE).
- **Chapter 2** provides an understanding of the organizations, capabilities, and training of CBRN operations. It is broken into two sections—CBRN organizations and capabilities and training.
- **Chapter 3** is the central idea of the publication. In this chapter the user can gain an understanding of the contributions of CBRN forces and their core functions are tied directly to supporting decisive action tasks of offense, defense and stability. It illustrates examples of how CBRN forces provide support to maneuver in and/or anticipated CBRN environments to maximize the effectiveness of combined arms operations and achieve freedom of action.

Based on current doctrinal changes, certain terms for which FM 3-11 is proponent have been rescinded or modified for purposes of this manual. The glossary contains acronyms and defined terms. See introductory table-1 and introductory table-2 for specific term changes.

Introductory table-1. Rescinded Army terms

Term	*Remarks*
chemical, biological, radiological, and nuclear consequence management	Army definition no longer used. Accept Joint characterization of chemical, biological, radiological, and nuclear response.
chemical, biological, radiological, and nuclear responders	Rescinded
chemical, biological, radiological, and nuclear threats	Rescinded
emergency management	Rescinded
weapons of mass destruction counterforce	Rescinded
weapons of mass destruction proliferation prevention	Rescinded

Introductory table-2. Modified Army terms

Term	*Remarks*
chemical, biological, radiological, and nuclear active defense	Adopt Department of Defense definition.
chemical, biological, radiological, and nuclear operations	Modified definition.

Introductory figure 1 depicts a logic chart to show the CBRN units and staffs contribution to unified land operations in anticipated CBRN OE. CBRN challenges in an OE include both peer threats and other considerations that may occur throughout the Army's strategic roles. The Army conducts unified land operations to support the joint force. CBRN units and staffs support unified land operations by conducting CBRN operations during decisive action, guided by mission command. The Chemical Corps provides tailorable, scalable, and adaptive CBRN reconnaissance, hazard mitigation, and expertise in support of decisive action to ensure freedom of action and survivability at home and abroad. CBRN forces are task-organized at echelon with leaders and Soldiers with the right tools and skills to support Army maneuver, counter and exploit the use of WMD, and provide layered and integrated protection from hazards. This is executed through the CBRN core functions (assess, protect, mitigate) and integrating activity of hazard awareness and understanding to enable protection and preserve the force's combat power to cope with CBRN hazards in the OE.

CBRN Challenges in an Operational Environment

Peer Threats

- Use of CBRN agents to fix, disrupt, isolate, block, defeat, and degrade friendly forces
- Integrated use of CBRN in hybrid and asymmetric warfare to shape operational environments and influence populations
- Proliferation of WMD materials, technology, and expertise
- Dense urban areas and subterranean environments
- Post WMD strike, collateral damage, and mass casualties
- Degradation of combat power

Other Considerations

- Increased sustainment requirements
- Isolated or cut lines of communication
- Atmospheric release of traditional or nontraditional agents or toxic industrial chemicals
- Escalation of hazard due to infectious pathogens
- Human factors (confusion and increased stress)
- Degradation of combat power
- Degraded communications
- Required integration with joint, interagency, multinational, and emergency responders

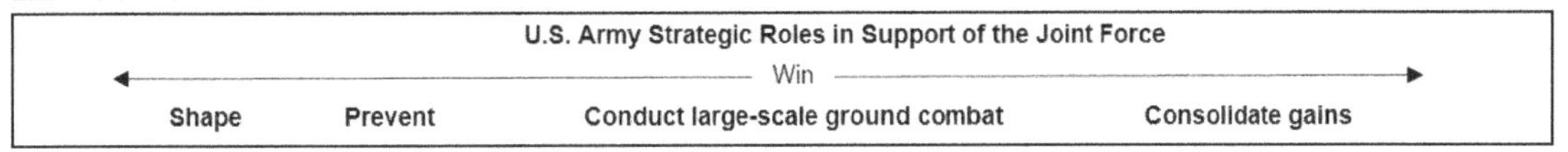

Unified Land Operations (The Army's Operational Concept)

Unified land operations are simultaneous execution of offense, defense, stability, and defense support of civil authorities across multiple domains to shape operational environments, prevent conflict, prevail in large-scale ground combat, and consolidate gains as part of unified action. (ADP 3-0)

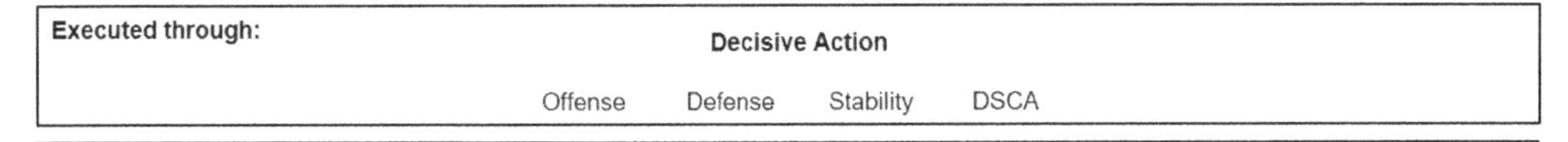

Guided by:

Mission Command
(Approach)

CBRN contributions to Army operations:

Provide tailorable, scalable, and adaptive CBRN reconnaissance, protection, hazard mitigation, and expertise in support of decisive action to ensure freedom of action and survivability at home and abroad.

Conducted at echelon by:

Through the integrating activity: **Hazard Awareness and Understanding**

Core functions:

Assess

- CBRN threats and hazards
- Critical asset/site vulnerabilities
- CBRN early warning systems
- Information collection and exploitation

Protect

- Freedom of action
- Force projection and staging
- Critical assets and sites
- High-value targets
- Lines of communication

Mitigate

- Enemy CBRN capabilities
- CBRN exposure to friendly forces
- Through decontamination operations
- Consolidate gains through countering WMD
- CBRN hazards through DSCA

Achieve through CBRN critical tasks:

- Understand CBRN threats and hazards in the environment
- Conduct CBRN information collection through reconnaissance and surveillance
- Conduct CBRN threat and hazard assessment
- Advise the commander on the impact of CBRN hazards on the mission
- Coordinate CBRN protection
- Conduct proactive, risk-based decision making
- Provide decontamination expertise and execution
- Destroy, dismantle, remove, transfer, and dispose or consolidate CBRN material
- Assess, characterize, and exploit CBRN hazards

Legend:

CBRN	chemical, biological, radiological, and nuclear	DSCA	defense support of civil authorities
		U.S.	United States
CBRNE	chemical, biological, radiological, nuclear, and explosives	WMD	weapons of mass destruction

Introductory figure 1. Logic chart

This page intentionally left blank.

Chapter 1
CBRN Operations Overview

This chapter describes how the CBRN units and staffs support large-scale combat operations and the associated challenges that Army forces currently encounter. It also addresses CBRN core functions, the anticipated OE, and the roles of CBRN units and staffs across the conflict continuum and Army strategic roles.

OVERVIEW

1-1. The current OE presents conditions that will challenge current and future commanders conducting unified land operations. The proliferation of WMD (and the constant pursuit of the materials, expertise, and technology required to employ WMD) will increase in the future. Through proxy forces, near-peer threats will operationalize emerging pathogens and new agents against civilian populations to increase confusion and inflict mass casualties. Neer-peer adversaries may create novel chemical warfare agents. These nontraditional agents may present unique challenges to CBRN defense capabilities and are expected to be employed for purposes that parallel those for traditional chemical agents if they are used on a smaller scale. The anticipated CBRN OE will contain dense urban environments and subterranean environments. Troop formations will contend with highly restrictive terrain because of CBRN agents employed to disrupt operations and because of industrial facilities that have the potential risk of causing exposure to agents or toxic industrial materials (TIMs). State and nonstate actors continue to develop WMD programs to gain advantage against the United States and its allies. The Army accomplishes its mission as part of a joint team to shape the OE, prevent conflict, conduct large-scale combat, and consolidate gains against a peer threat. To accomplish this mission, commanders must balance the inherent risks of military operations with mission accomplishment. This balance is achieved in part through protection tasks.

1-2. The likelihood of the enemy's use of WMD increases during large-scale combat operations—particularly against mission command nodes, massed formations, and critical infrastructure. Commanders ensure as much dispersion as is tactically prudent. In the offense, Army forces maneuver quickly along multiple axes, concentrate to mass effects, and then disperse to avoid becoming lucrative targets for WMD and enemy conventional fires. Commanders must anticipate that the high tempo formation will disintegrate or displace. Enemy formations may get intermingled with friendly forces or be bypassed, which requires follow-on and supporting units to protect themselves and dedicated forces to secure the consolidation area by defeating or destroying enemy remnants.

1-3. ***CBRN operations* is the employment of capabilities that assess, protect against, and mitigate the entire range of chemical, biological, radiological, and nuclear incidents to enable freedom of action.** CBRN operations support operational and strategic objectives to counter WMD and operate safely in a CBRN environment.

1-4. The capabilities provided by CBRN units and staffs support the Army operational concept—unified land operations. The implications of CBRN operations have a strategic and operational impact, even if the actions are tactical. Army actions in CBRN environments described in this field manual are linked to joint doctrine (JP 3-11, JP 3-40, and JP 3-41) and Army and multi-Service tactics, techniques, and procedures publications. The CBRN functions of assess, protect, and mitigate are the strength that CBRN units and staffs provide in support of operations in CBRN environments, CWMD, and CBRN response.

1-5. CBRN units and staffs must continue to evolve, providing the maneuver commander with dynamic options to address CBRN threats. In turn, CBRN staffs and formations provide the enduring confidence so that, if a CBRN hazard exists, it can be assessed, characterized, and exploited. CBRN staffs must also provide mitigation strategies, leveraging responsive applications, systems, and tactics.

1-6. One of the twelve primary protection tasks is conduct CBRN operations. This task includes the employment of tactical capabilities that anticipate and counter the entire range of CBRN threats and hazards. The protection warfighting function preserves the force so that commanders can apply maximum combat power to accomplish the mission. Commanders and staffs synchronize, integrate, and organize CBRN operations and resources with other protection capabilities to preserve combat power and identify and prevent or mitigate the effects of threats and hazards. Protection integrates all protection capabilities to safeguard the force, personnel (combatants and noncombatants), systems, and physical assets of the United States and its mission partners. Protection is not a linear activity—planning, preparing, executing, and assessing protection is a continuous and enduring activity.

Note. See ADP 3-37 for additional information on protection.

1-7. Effective CBRN operations require the full integration of CBRN Soldiers, units, and staffs as members of the combined arms team. CBRN forces integrated with maneuver forces contribute to a shared understanding of the OE and an integrated and synchronized approach to conducting operations in a complex CBRN environment. As Army professionals, our shared understanding and mutual trust are established and maintained through habitual training, persistent liaison, collaborative planning and preparation, standard operating procedures, clear command and support relationships, and effective mission rehearsals.

1-8. The following vignette describes the application of CBRN functions used by the Allied Expeditionary Force newly formed 1st Gas Regiment in World War I (WWI). The emergent use of chemicals to shape the battlefield impacted tactics and nearly changed the outcome of WWI.

Gas Warfare in the Meuse-Argonne Offensive

Over 47 days (from 26 September to the Armistice on 11 November 1918), the American Expeditionary Force was engaged in the largest battle yet fought in the history of the United States—the Meuse-Argonne Offensive. More than 1.2 million American Servicemen were committed to the battle in a combined allied effort to finally break through German lines and once again combat the enemy on an open battleground.

Throughout this campaign, toxic chemical agents first introduced on the battlefield in April 1915 were used effectively by German forces, inflicting a great number of casualties among American forces building up for the attack. From 1915 to 1918, the Germans held the initiative in most areas of gas warfare by introducing agents (such as phosgene and mustard gas) that allowed them to alter the tactical situation rapidly and by using gas to support maneuver during an infantry attack.

Initially, having no prior experience in addressing gas threats, the allies struggled to keep up with such offensive doctrine. The use of toxic chemicals by the enemy effectively hindered Allied forces freedom of maneuver, isolating forces from one another and denying key terrain. Areas saturated by chemicals could be impenetrable for days and disrupt operations. The key to the success of the campaign was the newly organized Chemical Warfare Service, which was tasked to provide offensive and defensive assistance to the American advance by providing gas training and smoke screens and by eliminating German machine gun positions with thermite. On 28 June 1918, the 1st Gas Regiment (also known as The Hellfire Boys) was formed. The tactical employment of the gas troops was to support the infantry before and during the battle.

Gas Warfare in the Meuse-Argonne Offensive (continued)

The 1st Gas Regiment ability to assess German intelligence and exploit captured German chemical warfare material was critical to understanding how the enemy could and would employ gas in future battles. They were able to rapidly develop new equipment and tactics to protect the force and preserve combat power for major operations. Lastly, their ability to provide decontamination within trench warfare allowed the 1st Gas Regiment to mitigate desired enemy effects on the terrain.

The success of the 1st Gas Regiment in support of American Expeditionary Forces highlights one of the very best examples of how CBRN forces support maneuver to win in a complex battlefield. Because of the 1st Gas Regiment ability to assess the enemy, protect the force, and mitigate the effects of gas warfare on the American Expeditionary Force, the German Army tactical advantage in using gas warfare was defeated.

CORE FUNCTIONS

Core Functions

Assess
Protect
Mitigate

Integrating Activity

Hazard Awareness and Understanding

1-9. The contributions of CBRN Soldiers are executed through the core functions of assess, protect and mitigate. A *function* is a practical grouping of tasks and systems (people, organizations, information, and processes) united by a common purpose (ADP 1-01). The CBRN functions communicate the CBRN tasks that provide the Army the means to accomplish its mission in a CBRN environment. These functions define the contribution of CBRN units and staffs to the Army and joint force. These core functions provide a focus for task and unit training, leader development, and force design. Hazard awareness and understanding is an integrating activity that links information obtained from all functions to better understand the OE. Understanding and excelling at these functions contribute to mission success in CBRN environments across all tasks of decisive action. These core functions may be executed individually, simultaneously, or sequentially. The CBRN core functions are assess threats and hazards, provide protection in and against CBRN environments, and mitigate CBRN incidents.

1-10. The functions can follow a process cycle, which allows them to feed into each other in either direction. Tasks conducted in the assess function provide necessary information for making proactive decisions. CBRN integrated into a reconnaissance and surveillance plan designed to provide early warning of a CBRN attack facilitates better assessment of enemy capabilities. Actions to mitigate CBRN incidents preserve future operations in the environment and provide insight into further assessments. All of these functions provide the basis for developing hazard awareness and understanding. Tasks within each function build on an initial awareness of hazards to further the commander's situational understanding of the environment and the impact of those hazards on current and future operations. Actions taken and information collected from assessing, protecting against, and mitigating CBRN incidents contribute to a better understanding of the OE, including CBRN hazards and their potential effects. The CBRN functions are described in the following paragraphs.

1-11. CBRN functions support CBRN defense tasks, including active and passive CBRN defense. *CBRN defense* is the measures taken to minimize or negate the vulnerabilities to, and/or effects of, a chemical, biological, radiological, or nuclear hazard or incident (JP 3-11). CBRN defense tasks are connected from joint doctrine through the multi-Service doctrine of passive defense and intersect with all of the CBRN functions. The combination of active and passive CBRN defensive measures reduces the effectiveness and success of CBRN weapon and improvised CBRN device employment and mitigates the risks associated with hazards.

Assess

1-12. Through information collection and dissemination, effective warning and reporting, modeling, and hazard awareness and understanding CBRN staffs and units provide the Army the ability to estimate the potential for (or the existence of) CBRN threats and hazards. Assessing hazards allows proactive decision making and encompasses all of the capabilities to evaluate the potential for CBRN threats and hazards in the OE, detect and model CBRN hazards, and determine the characteristics and parameters of hazards throughout the OE that bear on operational and tactical decisions. This function addresses the progression of CBRN capabilities—from just sensing hazards to avoid them to assessing hazards at a distance to enable the freedom of maneuver.

1-13. CBRN staffs provide commanders and planners with assessments of CBRN threats and hazards in the OE to integrate information from operations and intelligence. CBRN staffs provide the commander an evaluation of the risks and advise the commander in course of action (COA) development. Assessing hazards allows the commander to better understand the CBRN environment, assess the risk, and consider alternative options in the area of operations (AO). See ATP 3-11.36 for more information on staff actions that assess hazards.

1-14. At the tactical level, reconnaissance, surveillance, security, and intelligence operations are the primary means by which a commander conducts information collection to answer the commander's critical information requirements (CCIRs) to support decisive operations. (See FM 3-55 for more information.) A *chemical, biological, radiological, or nuclear incident* is any occurrence, resulting from the use of chemical, biological, radiological, and nuclear weapons and devices; the emergence of secondary hazards arising from friendly actions; or the release of toxic industrial materials or biological organisms and substances into the environment, involving the emergence of chemical, biological, radiological, and nuclear hazards (JP 3-11). Assessing CBRN hazards provides the foundation for an accurate and timely understanding of CBRN impacts on the OE. The tasks that are associated with this function are related to assessing and characterizing sites, reconnaissance and surveillance, and staff actions to provide assessments in the planning process.

1-15. Information collection tasks conducted during the planning and preparation phases of the operations process provide CBRN staffs information from the intelligence staff. The connection to the assistant chief of staff, intelligence (G-2)/battalion or brigade intelligence staff officer (S-2) and surgeon sections is necessary to feed information into threat assessments and aid in intelligence preparation of the battlefield (IPB). Information about CBRN threats and hazards in the OE helps the CBRN staff advise the commander so that he can assess and manage risk and consider which vulnerabilities to accept or mitigate. Primary tasks associated with this function include the following:

- Contribute to the IPB process.
- Conduct CBRN threat and hazard assessments.
- Provide operational and technical advice and planning recommendations on intelligence, surveillance and reconnaissance, and operations.
- Collect information on CBRN threats and hazards through reconnaissance and surveillance (R&S).
- Advise the commander on the impact of CBRN threats and hazards on the mission.
- Detect, locate, report, and mark hazards.
- Manage chemical, biological, and radiological survey and monitoring tasks.
- Estimate the effects of chemical, biological, and radiological exposure on mission accomplishment.

1-16. The Army intelligence process consists of four steps (plan and direct, collect and process, produce, and disseminate) and two enduring activities (analyze and assess). The assess threats and hazards function contributes to this process and uses products from it to further hazard awareness and understanding.

Protect

1-17. CBRN staffs and units provide the Army capabilities for protection against CBRN incidents. *Protection* is the preservation of the effectiveness and survivability of mission-related military and nonmilitary personnel, equipment, facilities, information, and infrastructure deployed or located within or outside the boundaries of a given operational area (JP 3-0). It encompasses the execution of physical defenses to negate the effects of CBRN hazards on personnel and material. Protection conserves the force by providing individual and collective protection postures and capabilities. Protecting the force from CBRN incidents includes hardening systems and facilities, preventing or reducing individual and collective exposures, or applying medical prophylaxes.

1-18. If the capability is available, an adversary may create CBRN hazards against mission command nodes, massed formations, or infrastructure to deny the freedom of action or key terrain during large-scale combat operations. Commanders balance the need to mass effects against the requirement to concentrate forces and ensure as much dispersal as is tactically prudent to avoid presenting lucrative targets for enemy fires and to mitigate the effects of CBRN incidents. Army units must train in CBRN defense to operate under CBRN conditions.

1-19. Tasks that support this function may occur throughout all phases of the operations. Many of the protect tasks for support to decisive action are described in chapter 3. Examples of protect tasks include the following:

- Train units to operate within CBRN environments.
- Employ assessments of unit capabilities and vulnerabilities.
- Protect personnel, equipment, and facilities from CBRN, including toxic industrial material (TIM) effects.
- Advise the commander on CBRN readiness.
- Coordinate the sustainment of CBRN defense equipment and medical chemical defense material.
- Employ proactive risk-based decision making.

MITIGATE

1-20. CBRN units and staffs provide the Army the ability to mitigate CBRN incidents by responding with the personnel, subject matter expertise, and equipment to reduce or neutralize the hazard. *Contamination mitigation* is described as the planning and actions taken to prepare for, respond to, and recover from contamination associated with all CBRN threats and hazards in order to continue military operations (JP 3-11). The mitigate function includes capabilities to negate hazards, such as the decontamination task.

1-21. Mitigating a CBRN incident encompasses a range of tasks to mitigate hazard effects after a CBRN incident. A CBRN incident can include deliberate attacks or accidental releases from technological or natural disasters. It includes all efforts to respond to CBRN incidents and reduce hazard effects on forces, populations, facilities, and equipment, including contamination mitigation and domestic and international CBRN response. Contamination mitigation contains two subsets—contamination control and decontamination—which are described in detail in ATP 3-11.32. The tasks conducted within domestic and international CBRN response are the same decontamination or hazard mitigation tasks conducted in tactical operations. The context in which they occur changes some of the operational considerations. These tasks are described in chapter 3. Tasks for mitigating CBRN incidents include the following:

- Provide a scalable response to CBRN incidents.
- Provide decontamination expertise and execution.
- Conduct WMD defeat, disablement, and/or disposal.
- Support health service support patient decontamination.
- Conduct modeling to determine the impact on operations.
- Perform incident and hazard response in support of defense support of civil authorities (DSCA).

HAZARD AWARENESS AND UNDERSTANDING

1-22. CBRN hazard awareness and understanding is a set of integrating activities happening at the individual and collective level to comprehend implications of CBRN environments on operations. Hazard awareness and understanding integrates all of the functions (assess, protect, mitigate), preincident through postincident, to facilitate situational understanding. Hazard awareness and understanding aids the CBRN staff in the collaborative process of IPB to provide the commander an understanding of how CBRN hazards in the AO affect mission accomplishment.

1-23. CBRN hazard awareness is achieved through the collection of data from individuals, sensors, or intelligence. CBRN hazard understanding is achieved through the fusion of all data sources of information. CBRN hazard understanding is the ability to comprehend the implication, character, nature, or subtleties of CBRN hazards and their impact on the OE, mission, and force to enable situational understanding. CBRN forces must be able to integrate all available information and translate technical information about the hazard into meaningful information that is useful to the commander in making risk decisions. Information collected from the execution of the tasks within the functions of assess, protect, and mitigate contributes to building upon hazard awareness to achieve understanding.

1-24. The tasks associated with hazard awareness and understanding may occur throughout all phases of operations. They are ongoing tasks that incorporate information obtained from the execution of all CBRN-related tasks.

CBRN CAPABILITIES ACROSS THE RANGE OF MILITARY OPERATIONS

1-25. The Army operates in a strategic environment that can result in military operations under many conditions. These operations are conducted within a ROMO—peace, conflict, and war. CBRN capabilities exist to support contingency and large-scale combat operations and to aid in DSCA, security cooperation, military engagement, and CWMD.

1-26. Force projection usually begins as a contingency operation or as a rapid response to a crisis. Contingency operations may be required for combat or noncombat situations. Contingency operations could be joint, interagency, intergovernmental, or multinational. Committed forces are tailored and task-organized for rapid deployment, effective employment, and mission accomplishment.

1-27. As WMD materials, technology, and expertise proliferate across the globe, it is likely that the United States will encounter them in military operations across the ROMO. Because CBRN threats and hazards make any operation more difficult, detailed planning is crucial. Many CBRN unit capabilities exist in the reserve component; therefore, the time required for mobilization must be considered. The amount and type of reserve forces that are mobilized depend on the crisis.

1-28. CBRN capabilities support operations conducted across the ROMO by assessing CBRN threats and hazards, providing protection against CBRN hazards, mitigating CBRN incidents, and providing hazard awareness and understanding. All CBRN functions share a common fundamental purpose that fits within the protection warfighting function to achieve or contribute to national objectives.

1-29. While the U.S. Army must be manned, equipped, and trained to operate across the range of military operations, large-scale ground combat against a peer threat represents the most significant readiness requirement. Chapter 3 provides an overview of how CBRN forces support combined arms operations through decisive action. See FM 3-0 for a discussion on large-scale ground combat operations.

ARMY STRATEGIC ROLES

1-30. The Army accomplishes its mission by supporting the joint force in four strategic roles: shape the OE, prevent conflict, conduct large-scale ground combat, and consolidate gains. Though this publication focuses primarily on large-scale ground combat operations and the consolidation of gains, CBRN support to shape the OE and prevent conflict are critical for the joint force to implement the national strategy and set desirable conditions to prevent and deter the adversary's undesirable actions.

Support to Shaping the Operational Environment

1-31. Army operations to shape the OE incorporate all activities intended to promote regional stability and set conditions for a favorable outcome in the event of a military confrontation. To prevent, dissuade, or deny adversaries or potential adversaries from possessing or proliferating WMD, forces must be prepared to support interdiction efforts, security cooperation, and nonproliferation efforts. These activities include support to planning, CWMD, targeting, and security cooperation.

Support to Planning

1-32. CBRN staffs support the development of campaign plans, operations plans, and concept plans through joint operational planning and the military decisionmaking process (MDMP). During planning, CBRN staffs recommend force structures that are tailored to assess, protect, and mitigate CBRN hazards. CBRN staffs also provide critical expertise to IPB to help evaluate the threat, determine possible threat courses of action, and describe the effects that potential CBRN hazards have on operations. CBRN planners provide insight into developing the CCIRs that shape information collection activities.

1-33. Logistical planning for CBRN incidents is conducted through running estimates that ensure that sufficient individual protective equipment, medical chemical defense material, decontamination kits, water, and CBRN-related consumables are available. Understanding the supply lead times needed to requisition CBRN supplies and distribute them rapidly to the force when required is critical to mission success. Logistician and CBRN staff coordination is essential to distribute CBRN equipment to forces as they flow into theater.

Support to Countering Weapons of Mass Destruction

1-34. *Countering weapons of mass destruction* are efforts against actors of concern to curtail the conceptualization, development, possession, proliferation, use, and effects of weapons of mass destruction, related expertise, materials, technologies, and means of delivery (JP 3-40). Army forces shape the OE by using cyber and space operations to dissuade or deter adversaries from developing, acquiring, proliferating, or using WMDs. In deliberate CWMD operations, commanders (with support from CBRN expertise) must confirm the existence of the site, characterization, and purpose of the facility. Tactically, CWMD is a combined arms mission that is task-organized and force-tailored to control, defeat, disable, and dispose of WMDs. Army forces must have an understanding of the threat to tailor a force capable of executing CWMD.

Support to Targeting

1-35. CBRN staffs provide subject matter expertise during the development of targets when the potential to encounter WMD sites exists. CBRN experts provide information on which materials can be found at the site, the threats posed by the materials at the site, and the impact of these threats on future operations. They also advise the targeting working group on the impact of the employment of WMD and the potential impact of targeted storage/production sites. CBRN staffs should understand enemy stockpile locations, potential transload locations, and delivery system methods and locations to enable deliberate and dynamic targeting. The CBRN staff collaborate with the civil affairs and medical staff to develop messaging for the civilian population on the implication of WMD in the AO.

Security Cooperation

1-36. Security cooperation builds security relationships that promote specific U.S. security interests, develop allied and friendly capabilities, and provide U.S. forces with peacetime and contingency access to partner nations. Security cooperation allows the transfer of technology and know-how to partner nations and allows direct observation and interaction to ensure that equipment and training are used properly.

1-37. CBRN forces participate in combined arms exercises and training, along with national assistance efforts (to include security assistance and foreign internal defense) to improve partnering and cooperation. Additionally, it fosters tactical CBRN interoperability between the United States and other partner nation forces.

Support to Prevent Conflict

1-38. Army operations to prevent include all activities meant to deter an adversary's undesirable actions and protect friendly forces as they deploy into theater. Army protection capabilities support operations to prevent during mobilization, the transit of Army forces and cargo, initial staging areas, and subsequent assembly areas (AAs) where uncertain threat conditions require a delicate balance between protection and building combat power.

1-39. Operations to prevent create the conditions required to quickly transition into large-scale ground combat. The challenge lies in being able to quickly provide specific CBRN forces to support an operation. The majority of the CBRN force resides in the reserve forces, requiring significant consideration of the order of deployment for force packages. A critical planning consideration requires the correct CBRN capabilities required to meet the combatant commander's (COCOM's) priorities to provide credible deterrence that defeats enemy considerations to employ CBRN in competition below armed conflict.

1-40. As part of a combined arms team, CBRN forces deter aggression and CBRN usage by actions that may disrupt an adversary's plans to employ them. Deterrence of CBRN usage is accomplished through visible displays of readiness and an advanced capability to operate in a CBRN environment. Commanders maximize readiness exercises to demonstrate a robust CBRN scheme of protection that deters undesirable actions involving CBRN.

OPERATIONAL ENVIRONMENT

1-41. The *operational environment* is a composite of the conditions, circumstances, and influences that affect the employment of capabilities and bear on the decisions of the commander (JP 3-0). Commanders and leaders charged with conducting operations in a CBRN environment must begin with a thorough understanding of the OE, the risks and opportunities associated with the OE, and the ways and means available for preserving combat power through protection. Leaders analyze the OE through the operational variables (political, military, economic, social, information, infrastructure, physical environment, and time [PMESII-PT]) and mission variables (mission, enemy, terrain and weather, troops and support available, time available, and civil considerations [METT-TC]) to provide an understanding that helps identify current, developing, and potential hazards and threats and enable the tasks to mitigate or eliminate them. Through continuous analysis of the OE, civil affairs staff can provide analyzed and evaluated civil considerations data concerning the host nation and indigenous populations and institutions. Military and civil capabilities include CBRN defense and decontamination capabilities; general information regarding the existence or movement of CBRN materials; local expertise, resources, or technology related to CBRN operations; and industrial CBRN processing, storage, or experimental facilities.

Note. See ATP 2-01.3 and ATP 3-11.36 for more information about analyzing the OE through PMESII-PT and METT-TC. See FM 3-57 for more information on civil considerations.

1-42. CBRN survivability is divided into chemical, biological, and radiological survivability, which is concerned with chemical, biological, and radiological contamination, including fallout. Also included is nuclear survivability, which covers initial nuclear weapon effects, including blast, electromagnetic pulse, and other initial radiation and shockwave effects. DoDI 3150.09 establishes policies and procedures for ensuring the survivability of the force to operate in chemical, biological, and radiological or nuclear environments as a deterrent to adversary use of WMDs against the United States, its allies, and interests. The ability of the force to operate in these environments must be known and assessed on a regular basis according to the nature of the OE. *Chemical, biological, radiological, nuclear, and explosives* are components that are threats or potential hazards with adverse effects in the operational environment (ATP 3-37.11).

1-43. Large urban areas and complex subterranean environments continue to present a concern for CBRN operations. Dense populations create concerns for pandemic diseases, the rapid spread of infections, mass casualties, and chaos created by panicked individuals. Operations in subterranean environments have unique hazards, such as poor air circulation from insufficient ventilation or a lack of breathable air due to toxic vapor displacement. As listed in ATP 3-21.51, subterranean environments exist in three major categories—tunnels or natural cavities and caves, urban subsurface systems, and underground facilities. The production and storage of CBRN materials and WMD (which may occur in subterranean environments and can increase atmospheric hazards), requires the alignment of CBRN enablers with maneuver forces. This combined arms effort allows forces to successfully manage, plan/account for, isolate, clear, exploit, and transition underground facilities.

CBRN Threats and Hazards

1-44. Threats and hazards have the potential to cause personal injury, illness, or death; equipment or property damage or loss; or mission degradation. CBRN threats include chemical, biological, or nuclear weapons; WMD programs; and improvised CBRN devices that produce CBRN hazards. When an improvised explosive device also utilizes a CBRN hazard to produce effects, it becomes an improvised chemical device, improvised biological device, improvised radiological device, or improvised nuclear device. A brief overview of the considerations associated with each of the C-B-R and N threats and hazards is important to fully grasp the influences each has on the decisions of the commander.

Note. Technical information about these hazards can be found in TM 3-11.91.

1-45. CBRN threats include the intent and capability to employ weapons or improvised devices to produce CBRN hazards. In contrast, CBRN hazards include CBRN material created from accidental or deliberate releases of TIMs, chemical and biological agents, nuclear materials, radiological materials, and those hazards resulting from the employment of WMD or encountered by the U.S. armed forces during the execution of military operations.

1-46. Commanders and their staffs must consider CBRN threats and hazards during integrating processes (IPB, targeting, risk management) and continuing activities (liaison, information collection, security operations, protection, terrain management, airspace control). IPB provides CBRN planners with intelligence regarding CBRN threats and hazards that impact freedom of maneuver. The continuing activity of information collection should direct reconnaissance and surveillance toward confirming CBRN-specific priority intelligence requirements (PIRs). These actions are critical to the CBRN integrating activity of hazard awareness and understanding, which supports the maneuver commander's ability to seize, retain, and exploit the initiative to maintain a relative position of advantage.

Note. See ADP 5-0 for additional information on integrating processes and operational or mission variables.

Threat

1-47. A *threat* is any combination of actors, entities, or forces that have the capability and intent to harm U.S. forces, U.S. national interests, or the homeland (ADRP 3-0). The use of CBRN weapons by the threat can have an enormous impact on all combat actions. Chemical employment is more likely than biological, radiological, or nuclear types because the threat can predict the effects of chemical agents and then use them to affect limited areas of the battlefield. The effects of chemical weapons are more predictable; therefore, they are more readily integrated into battle plans at the tactical level. The threat may also have biological, nuclear, and radiological capabilities that deserve consideration, despite the lower probability of employment. This is generally true of nuclear and biological weapons, which have lethal effects over much larger areas than do chemical weapons. The effects of biological weapons can be difficult to localize and employ in combat without them also affecting threat forces.

1-48. In response to foreign development, peer or near-peer threats maintain the capability to conduct chemical, nuclear, and possibly biological or radiological warfare. However, they initially prefer to avoid the use of CBRN weapons to prevent an international or multilateral response. Force modernization has introduced a degree of flexibility that was previously unavailable to combined arms commanders. It creates multiple options for the employment of forces at strategic, operational, and tactical levels with or without the use of CBRN weapons.

1-49. Many of the same delivery means available for CBRN weapons can also be used to deliver precision weapons that can often achieve desired effects without the stigma associated with CBRN weapons. The threat might use CBRN weapons to deter aggression or as a response to an enemy attack. It could use CBRN against a neighbor as a warning to a potential adversary to let the adversary know that it is willing to use such a weapon. It may use, or threaten to use, CBRN weapons that have collateral effects on noncombatants as a way of applying political, economic, or psychological pressure to show its determination and scope of aggression. Peer or near-peer threats have surface-to-surface missiles that are capable of carrying chemical, biological, or nuclear warheads. (Refer to the Worldwide Equipment Guide for specific unclassified threat weapon capabilities.) Additionally, a peer threat could use aircraft systems and cruise missiles to deliver a CBRN attack. They also maintain the capability of using special-purpose forces as an alternate means of delivering CBRN munitions packages. It is important to note the increased likelihood of peer threats employing tactical nuclear capabilities when there is a perceived threat to their sovereign borders or government regime.

1-50. The likelihood of the threat use of CBRN weapons increases during large-scale combat operations, particularly in situations where the primary objective is to break contact to prevent destruction or defeat. The use of CBRN weapons and the constant pursuit of the materials, expertise, and technology to employ them will increase in the future. State and nonstate actors continue to develop programs to gain advantage against the United States and its allies.

1-51. Threat forces determine suitable targets for CBRN weapons based on their perception of friendly vulnerabilities on that kind of attack and how the attack achieves their desired effects. The threat considers the following targets to be suitable for the employment of CBRN weapons:

- Precision weapons.
- Prepared defensive positions.
- Reserve and troop concentrations.
- Communication centers.
- Reconnaissance, intelligence, surveillance, and target acquisition centers.
- Key air defense sites.
- Logistics installations, especially port facilities.
- Airfields not intended for immediate future use.

Hazard

1-52. A *hazard* is a condition with the potential to cause injury, illness, or death of personnel; damage to or loss of equipment or property; or mission degradation (JP 3-33). CBRN hazards are CBRN elements that can create adverse effects due to an accidental or deliberate release and dissemination. Understanding hazards also helps the commander visualize potential impacts on operations. CBRN hazards create conditions that can damage or destroy life or vital resources or prevent mission accomplishment. CBRN hazards include toxic industrial chemicals (TICs), toxic industrial biologicals (TIB), toxic industrial radiologicals, and special nuclear materials collectively known as TIM.

Chemical Hazards

1-53. The types of chemical hazards that are of concern to the military have expanded tremendously over the last decade and now include a large number of TIC. Chemical hazards are any chemicals (manufactured, used, transported, or stored) that can cause death or other harm through the properties of those materials. Adversaries have the potential to employ these hazards to limit the friendly scheme of maneuver.

1-54. Chemical hazards can be divided into the following categories:

- **Chemical warfare agents.** These are specific agents (such as nerve or blister) that are developed as military weapons and designed to kill or severely incapacitate personnel and disrupt movement. Chemical warfare agents are generally considered lethal.
- **Military chemical compounds (other than chemical warfare agents).** These are chemical compounds that are developed, in part, for military use (riot control agents, smokes, obscurants), but not as weapons. Military chemical compounds are generally considered nonlethal and cause temporary illness. Toxic properties are primarily associated with improper use.
- **TICs.** These are widely available, commercial, natural, or manufactured compounds that pose a risk of direct and immediate harm to personnel. These same chemicals are considered plausible candidates for terrorist activities. In general, the risk of a TIC being used as a weapon depends on the severity of effects that it may cause and the probability that it may be obtained and released in dangerous quantities.

1-55. A chemical agent is a chemical substance that is intended for use in military operations to kill, seriously injure, or incapacitate, mainly through physiological effects. The term excludes riot control agents when used for law enforcement purposes, herbicides, smoke, and flame. Chemical agents are classified according to the following:

- **Physical state**. Agents may exist as a solid, liquid, or vapor.
- **Physiological action.** Based on their physiological effects, there are nerve, blood, blister, choking, and incapacitating agents.
- **Use.** The terms persistent and nonpersistent describe the time that an agent stays in an area. An adversary may have to expend large quantities of chemical agents to cause mass casualties or achieve area denial.
  - **Persistent agent.** A *persistent agent* is a chemical agent that, when released, remains able to cause casualties for more than 24 hours to several days or weeks (JP 3-11).
  - **Nonpersistent agent.** A *nonpersistent agent* is a chemical agent that when released dissipates and/or loses its ability to cause casualties after 10 to 15 minutes (JP 3-11).

1-56. Nontraditional chemical agents are a broad group of chemicals that fall outside the traditional chemical agent categories. These agents are not included in the Chemical Weapons Convention schedules, which were designed to capture traditional chemical warfare agents. While nontraditional agents possess some of the properties of traditional chemical agents, these properties often present unique challenges. Some nontraditional agents include lethal and incapacitating agents.

Biological Hazards

1-57. A *biological agent* is a microorganism (or a toxin derived from it) that causes disease in personnel, plants, or animals or causes the deterioration of materiel (JP 3-11). Biological agents are microorganisms that are capable of spreading disease through humans and agriculture (plants and animals). Biological agents are dispersed or employed as pathogens or toxins that cause disease in personnel, animals, and plants.

- **Pathogens.** Pathogens are disease-producing microorganisms (bacteria, viruses, fungi) that directly attack human, plant, or animal tissue and biological processes.
- **Toxins.** Toxins are poisonous substances that are produced naturally by bacteria, plants, fungi, snakes, insects, and other living organisms and may also be produced synthetically. Naturally occurring toxins are nonliving byproducts of cellular processes that can be lethal or highly incapacitating.

1-58. Pathogens require an incubation period to establish themselves in the body of a host and produce disease symptoms. The onset of visible symptoms may occur days or weeks after exposure. Some toxins can remain active for extended periods in the natural environment. This stability creates a persistent transfer hazard. Unlike chemical, radiological, and nuclear hazards, biological hazards are unpredictable, and it is difficult to classify the extent of the hazard.

1-59. Biological hazards in the form of TIBs include infectious agents and other biological hazards. They are often generated as infectious waste, such as on sharp-edged medical instruments (needles, syringes, and lancets) and material contaminated by bodily fluids.

1-60. Biological hazards can provide important advantages to adversaries who use them because of factors such as—

- Easy clandestine employment.
- Delayed onsets of symptoms.
- Detection, identification, and verification difficulties.
- Agent persistence.
- Communicability. These factors, combined with the factors listed below, increase the threat of biological agents:
  - Small doses can produce lethal or incapacitating effects over an extensive area.
  - They are difficult to detect in a timely manner.
  - They are easy to conceal.

- They can be covertly deployed.
- The variety of potential biological agents significantly complicates effective preventative or protective treatment.

Radiological Hazards

1-61. Radiological hazards include any nuclear radiation (such as electromagnetic or particulate radiation) that is capable of producing ions that cause damage, injury, or destruction. The Army is responsible for enforcing precautions and establishing tactics, techniques, and procedures for handling conventional munitions that employ radioactive materials, such as depleted uranium. This includes enforcing standards that protect personnel against alpha particle inhalation and ingestion and external beta, gamma, and neutron exposure. In addition, dangerous levels of radiation can result from damaged industrial radiation hazard areas. Due to the downwind hazards that such damage can produce, avoidance is the most effective individual and unit protective measure against industrial radiation hazards.

1-62. Radiological agents are a result from the decay of radioactive material by fission and can be classified as alpha particles, beta particles, neutrons, gamma particles, or X-rays.

- **Alpha particles.** Alpha particles are positively charged, highly energetic nuclei (two protons, two neutrons, but no electrons) that travel through air only a few inches from the nuclei that emit them. Alpha particles are easily stopped by a piece of paper or human skin.
- **Beta particles.** Beta particles are electrons or positrons that are ejected from the nucleus of an unstable atom. They travel further than alpha particles and are stopped by low-density materials such as aluminum, plastic, or tightly woven fabric (such as Joint Service Lightweight Integrated Suit Technology).
- **Neutrons.** Neutrons originate in the nucleus of an atom. They have no charge, but they do have a substantial mass, travel long distances, and are slowed and absorbed by a significant portion of hydrogenated material (water and plastics).
- **Gamma particles and X-rays.** Gamma particles and X-rays are electromagnetic energy with no mass and no charge. They travel long distances and are absorbed by high density materials.

Note. See TM 3-11.91/MCRP 10-10E.4/NTRP 3-11.32 for more information.

Radiological Devices

1-63. A *radiological dispersal device* is an improvised assembly or process, other than a nuclear explosive device, that is designed to disseminate radioactive material in order to cause destruction, damage, or injury (JP 3-11). In a radiological dispersal device, conventional explosives are bundled with radioactive materials.

1-64. A *radiological exposure device* is a radioactive source that is placed to cause injury or death (JP 3-11). A radiological exposure device may be hidden or buried where the penetrating radiation source within it (gamma and/or neutron) can affect the intended target. If it remains undetected, the potential dose to the intended target increases.

1-65. Adversaries can disperse radioactive material in many ways; for example, they can use a conventional platform to deliver radioactive materials that may be obtained from industrial sources, such as radioactive material from a power-generating nuclear reactor or from use in industry, medicine, or research. Unless radioactive sources are thoroughly shielded, improvised devices employing these materials will likely have a significant radiological signature that can be detected before detonation, dispersal, or deployment. The dispersal of radioactive material represents an inexpensive capability that requires limited resources and technical knowledge.

Nuclear Hazards

1-66. The severity of nuclear hazards depends on the weapon yield, emission spectrum, and distance to the target. Severity also depends on the height of burst, especially for fallout production and electromagnetic pulse. Residual radiation effects are due to emissions (typically alphas, betas, and low-energy gammas) from fission fragments (atoms produced during fission) and activated environmental materials (material that absorbs radiation and become radioactive itself). Collectively, these sources are called fallout.

1-67. Nuclear hazards are the applied effects on personnel, equipment, units, and systems. Nuclear hazards are produced by the energy released from a nuclear weapon employed offensively or defensively. When detonated, a typical nuclear weapon releases its energy as blast, thermal radiation (including X-rays), nuclear radiation (alpha and beta particles, gamma rays, and neutrons), and electromagnetic pulse.

1-68. Nuclear weapon effects are qualitatively different from biological or chemical weapon effects. The nature and intensity of nuclear detonation effects are determined by the type of weapon, its yield, and the physical medium in which the detonation occurs. Weather conditions affect fallout immensely. Some characteristics of nuclear weapon effects include the following:

- The distribution of energy and the relative effects of blast, heat, and radiation depend largely on the weapon and the altitude at which it is detonated.
- A typical nuclear weapon releases most of its energy as thermally generated X-rays at the point of detonation.
- The amount of fallout depends on the weapon yield, weapon type, and height of the burst.
- The area affected depends heavily on the wind.
- Surface bursts produce the most fallout.
- The hazard to personnel depends on the level of radiation present and the duration of exposure.

Note. See TM 3-11.32 for more information on how to understand and mitigate nuclear hazards.

OPERATIONAL FRAMEWORK

1-69. The operational framework provides a tool for visualizing and describing operations by echelon in time and space within the context of an AO, area of influence, and area of interest. The operational framework has the following four components:

- Area of operations.
- Deep, close, support, and consolidation areas to describe the physical arrangement of forces.
- Decisive, shaping, and sustaining operations to further articulate an operation in time.
- Main and supporting efforts.

1-70. Figure 1-1, page 1-14, is an example to help visualize the use of CBRN forces in a contiguous corps AO. In the close area, CBRN reconnaissance forces may conduct mounted reconnaissance or hazard site assessments during decisive operations. A CBRN battalion may support the division by using assigned companies that have decontamination and reconnaissance assets to enable decisive, shaping, or sustaining operations. CBRN battalions may also task-organize hazard response companies to brigade combat teams (BCTs). WMD coordination teams may be assigned to division through theater headquarters as needed. The CBRN staff elements that support the framework at every echelon during all operations are not depicted in the graphic. Area support and biological companies may support the joint security area with decontamination and biosurveillance to protect the force.

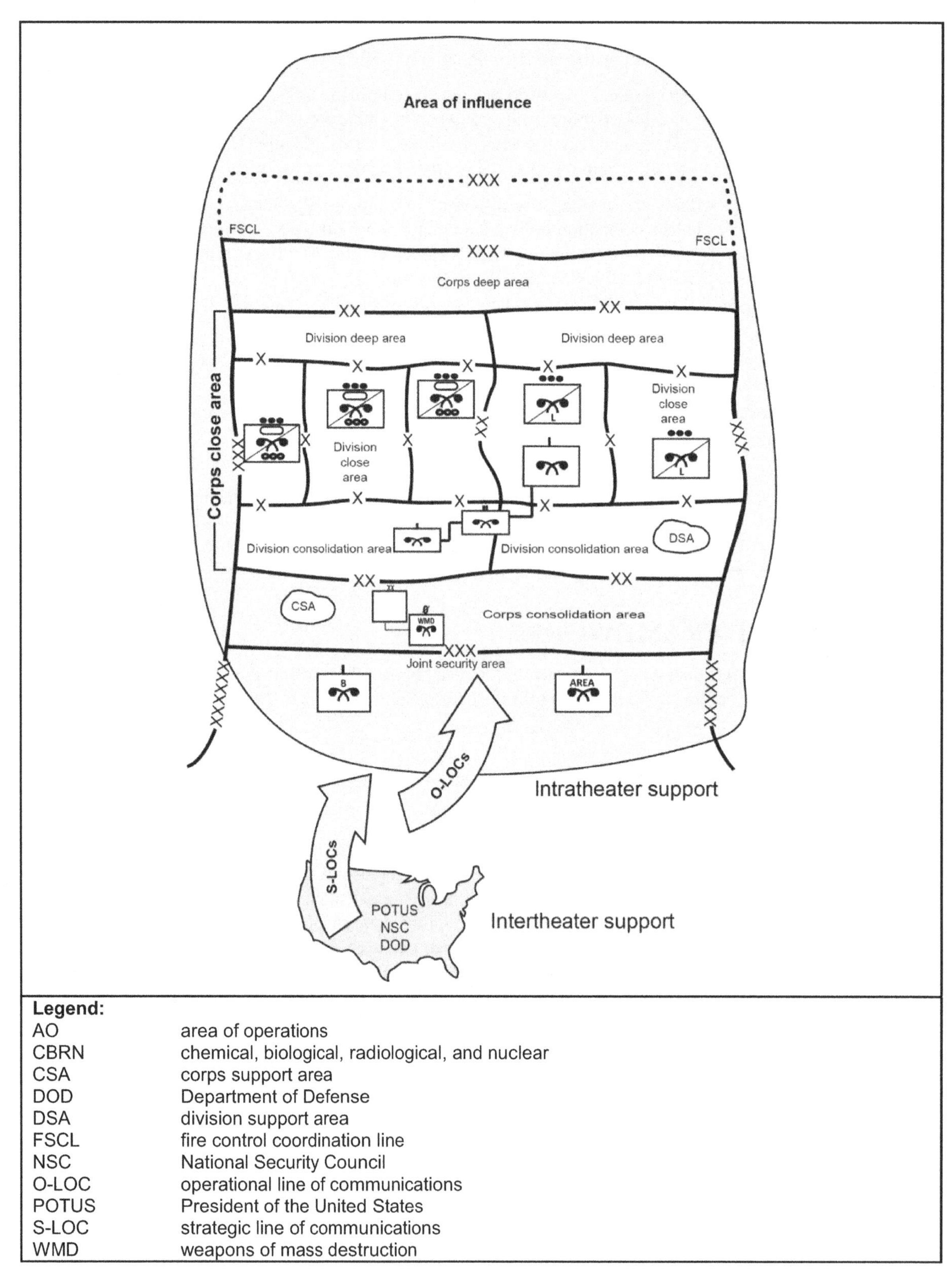

Legend:

AO	area of operations
CBRN	chemical, biological, radiological, and nuclear
CSA	corps support area
DOD	Department of Defense
DSA	division support area
FSCL	fire control coordination line
NSC	National Security Council
O-LOC	operational line of communications
POTUS	President of the United States
S-LOC	strategic line of communications
WMD	weapons of mass destruction

Figure 1-1. Example CBRN forces in contiguous corps AO

Chapter 2

CBRN Organizations, Capabilities, and Training

This chapter is divided into two sections. CBRN units and staffs provide a mix of expertise and capabilities to commanders in support of joint and Army operations in a complex CBRN environment. To provide these capabilities across the Army, the Chemical Corps provides an echeloned array of CBRN Soldiers, staffs, and units that can integrate with maneuver forces and provide the necessary expertise and capabilities. Section I provides information for commanders and staffs at all echelons to understand CBRN organizations, their capabilities, and their limitations. Section II provides information on CBRN training.

SECTION I—CBRN ORGANIZATIONS AND CAPABILITIES

2-1. CBRN staffs and units are organized into scalable, tailorable, and multifunctional formations that can best support joint and Army operations. The tailorable force increases the mission command and sustainment capability for battalions and is responsive to and aligned with BCTs, divisions, and corps.

2-2. CBRN staffs are limited in size and capacity across all echelons. CBRN operations are not the sole responsibility of the CBRN staff, but they must be integrated and executed across staff sections. When CBRN incidents occur, logistics, medical support, maintenance support, and every staff section is impacted and quickly overwhelmed if they are not prepared collectively to execute large-scale combat operations under CBRN conditions.

2-3. The balance of actions taken by CBRN platoons, companies, and battalions falls into the functions of assess threats and hazards and mitigate CBRN incidents. CBRN units are primarily tasked with assess tasks (such as reconnaissance and surveillance) and mitigate tasks (such as decontamination). However, it is a leader's responsibility to ensure that all Soldiers train and execute tasks which support assessing, protecting against, and mitigating CBRN hazards.

2-4. Table 2-1, page 2-2, provides a matrix of the capabilities for conducting CBRN operations that exist at different echelons and units. See ATP 3-11.36 and ATP 3-05.11 for detailed information about CBRN staffs, organizations, and detachments across the Army and other Services. The Xs in table 2-1 represent the best-employment capabilities. Many of the units have the capability to complete several of the tasks listed; however, they may not be the best-resourced or possess the capability that other units possess.

Table 2-1. CBRN capabilities matrix

Army Component	Higher Headquarters	CBRN Unit	Advise on CBRN hazards	Detect and provide field confirmatory identification of known CBRN hazards	Provide dismounted assessments	Provide early warning of contamination (contamination avoidance)	Report, mark, and identify bypass routes around contamination	Collect and transfer samples	Assess hazards in support of site exploitation and CBRN response	Detect biological warfare agent employment as a measure to provide medical treatment	Perform CBRN sample collection and management	Perform sensitive-site assessment and characterization	Perform munitions assessment and disablement	Perform specialized sampling support to forensics	Conduct CBRN survey to determine the nature, scope, and extent of CBRN hazard	Conduct decontamination tasks	Conduct terrain decontamination	Conduct aircraft decontamination	Conduct fixed site decontamination	Conduct sensitive-site exploitation	Nonintrusively assess chemical munitions	Perform technical escort of CBRN material	Conduct laboratory analysis to provide up to theater level identification
A/NG	BCT	Recon PLT (IBCT)	X	X	X	X	X	X	X		X	X											
A/NG	BCT	Recon PLT (ABCT)	X	X		X	X	X	X	X	X												
A/NG	BCT	Recon PLT (SBCT)	X	X		X	X	X	X	X	X												
A/AR/NG	Hazard Response Company	Hazard Assessment Platoon	X	X	X		X	X	X		X	X		X		X		X					
A/AR/NG	Hazard Response Company	R&S PLT (NBCRV)	X	X		X	X	X	X	X	X												
A	CBRNE Response Company	CBRNE Response Tm (CRT)	X	X	X			X	X		X	X	X	X	X					X	X	X	
AR/NG	Area Support Company	Recon PLT (Light)	X	X	X	X	X	X	X		X	X											
AR/NG	Area Support Company	Decon PLT (Heavy)	X						X							X	X	X	X				
AR/NG	Area Support Company	CBRN PLT (Bio)	X 1	X 1				X 1	X 1	X	X 1												
A/NG	Special Operations Forces	CBRN Recon Detachment (CRD)	X	X	X	X	X	X	X	X	X	X		X	X					X		X	
A/NG	Special Operations Forces	CBRN Decon Detachment (CDD)	X					X	X		X					X							

Table 2-1. CBRN capabilities matrix (continued)

Army Component	Higher Headquarters	CBRN Unit	Advise on CBRN hazards	Detect and provide field confirmatory identification of known CBRN hazards	Provide dismounted assessments	Provide early warning of contamination (contamination avoidance)	Report, mark, and identify bypass routes around contamination	Collect and transfer samples	Assess hazards in support of site exploitation and CBRN response	Detect biological warfare agent employment as a measure to provide medical treatment	Perform CBRN sample collection and management	Perform sensitive-site assessment and characterization	Perform munitions assessment and disablement	Perform specialized sampling support to forensics	Conduct CBRN survey to determine the nature, scope, and extent of CBRN hazard	Conduct decontamination tasks	Conduct terrain decontamination	Conduct aircraft decontamination	Conduct fixed site decontamination	Conduct sensitive-site exploitation	Nonintrusively assess chemical munitions	Perform technical escort of CBRN material	Conduct laboratory analysis to provide up to theater level id
A	Special Operations Forces	Decon & Recon Team (DRT)	X					X	X		X									X		X	
A	CBRNE CMD	Nuclear Disablement Team (NDT)	X								X 2									X 2			X
A	CBRNE CMD	WMD Coordination Team (WCT)	X																				
A	CBRNE CMD	CARA Light & Heavy Expeditionary Labs	X																				X
A	CBRNE CMD	CARA Remediation Response Team (RRT)	X																	X		X	
A / AR / NG	All	CBRN Staff	X																				

Legend:

A	Army	CMD	command
ABCT	armored brigade combat team	IBCT	infantry brigade combat team
BCT	brigade combat team	labs	laboratories
Bio	biological	NBCRV	nuclear, biological, and chemical reconnaissance vehicle
CARA	chemical, biological, radiological, nuclear, and high-yield explosives analytical and remediation activity	NG	National Guard
		PLT	platoon
CBRN	chemical, biological, radiological, and nuclear	recon	reconnaissance
		R&S	reconnaissance and surveillance
CBRNE	chemical, biological, radiological, nuclear, and explosives	SBCT	Stryker brigade combat team
		Tm	team

2-5. Figure 2-1 describes an example for the alignment of CBRN assets across two corps and against a peer threat. This example includes units from multiple components. A CBRN brigade that is supported by two CBRN battalions and that is composed of five chemical, biological, radiological, nuclear, and explosives (CBRNE) companies, an area support company, and a WMD coordination team, would be at the geographic/theater combatant command level collocated with the theater command. Area support companies are located at seaports of debarkation and aerial ports of debarkation to provide assessment and mitigation for joint reception, staging, and onward integration support. A special forces chemical reconnaissance detachment within the special forces group provides support for CWMD operations. In the corps AO, a CBRN brigade or battalion is assigned with a mix of CBRN units to support the consolidation of gains. In the division AO, each division has a CBRN battalion in direct support with a hazard response and biosurveillance company to provide an enhanced assessment and mitigation capability to close and consolidation areas. Each BCT has a hazard response company that is in direct support to it, increasing the capability to assess and mitigate CBRN hazards at tempo.

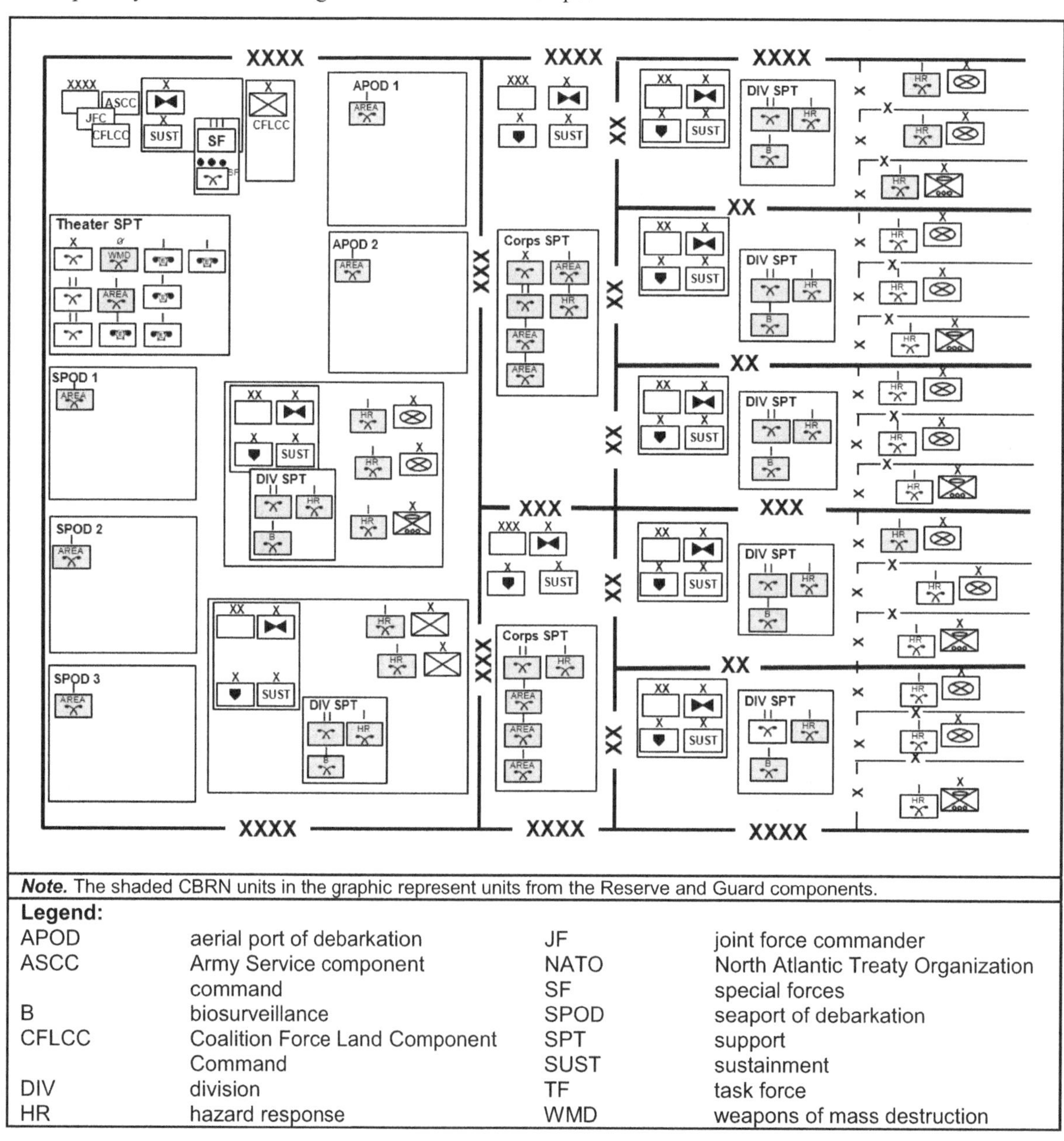

Note. The shaded CBRN units in the graphic represent units from the Reserve and Guard components.

Legend:

APOD	aerial port of debarkation	JF	joint force commander
ASCC	Army Service component command	NATO	North Atlantic Treaty Organization
		SF	special forces
B	biosurveillance	SPOD	seaport of debarkation
CFLCC	Coalition Force Land Component Command	SPT	support
		SUST	sustainment
DIV	division	TF	task force
HR	hazard response	WMD	weapons of mass destruction

Figure 2-1. Force illustration from theater to BCT

CBRNE COMMAND

2-6. The Department of the Army organized the CBRNE command as a full-spectrum, deployable, operational-level command to manage existing CBRN and explosive ordinance disposal (EOD) response assets. The CBRNE command is a JTF-capable headquarters that is capable of deployment in support of a wide range of CWMD and CBRNE activities. The CBRNE command integrates, coordinates, deploys, and provides trained and ready CBRN and EOD response forces. It exercises mission command of CBRN and EOD forces in support of joint and Army force commanders. When the CBRNE command operational command post is deployed, it is capable of integrating with the supported command headquarters staff or of operating as a separate element to conduct CBRN, EOD, and CWMD planning and has the capacity to execute simultaneous missions within and outside the continental United States across unified land operations. The CBRNE command maintains technical links with the appropriate Joint, federal, and state CBRN and EOD assets and with research, development, and technical communities to assure response readiness. See figure 2-2.

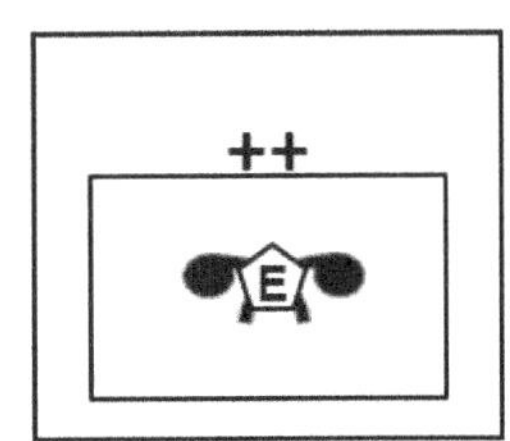

Figure 2-2. CBRNE command graphic

2-7. The CBRNE command can provide support through specialized teams: WMD coordination teams; nuclear disablement teams; chemical, biological, radiological, nuclear and high-yield explosives analytical and remediation activity (CARA); and the area medical laboratory (AML). WMD coordination teams are provided through the CBRNE command to supported and supporting headquarters at the division level and above to allow liaison and advice across the spectrum of CBRNE threats. WMD coordination teams provide specialized CBRNE staff augmentation and technical subject matter expertise to support planning and coordination for countering CBRNE and WMD threats in support of operational and theater requirements. The teams are composed of deployable CBRN, EOD, and nuclear experts who have organic intelligence and communications assets and can advise supported divisions, corps, Army Service component commands, or other organizations on CBRNE planning and operations.

2-8. The CBRNE command is supported by two deployable field laboratory elements—the AML and CARA mobile expeditionary laboratories. Both mobile expeditionary laboratories provide advanced technologies for theater validation identification. The CARA maintains remediation response teams to conduct the assessment and remediation of recovered chemical warfare material in the CONUS and OCONUS. The AML has a robust diagnostic capability to detect and identify a wide range of environmental contaminants, such as chemical/biological contaminants and/or radioisotopes. When deployed, the AML serves the COCOM as a theater level asset.

2-9. Nuclear disablement teams are the only DOD organizations capable of assessing, exploiting, characterizing, and disabling facilities associated with the nuclear fuel cycle in semipermissive or permissive environments. They advise commanders on the risks associated with these facilities, provide detailed information related to potential material proliferation, and can make recommendations on how to dispose of nuclear material.

Note. See ATP 3-37.11 and FM 3-94 for more information on the CBRNE command.

BRIGADE

2-10. CBRN brigades exist within all three components (Regular Army, U.S. Army Reserve, and Army National Guard) and provide command and control for battalions and separate companies that have a wide range of capabilities, including CBRN reconnaissance, decontamination, and biological detection. A CBRN brigade is dependent on other elements for administrative, logistic, medical, and maintenance support. See figure 2-3.

> ***Note.*** More information about the various battalion and company capabilities can be found in paragraph 2-18, paragraph 2-25, and ATP 3-11.36.

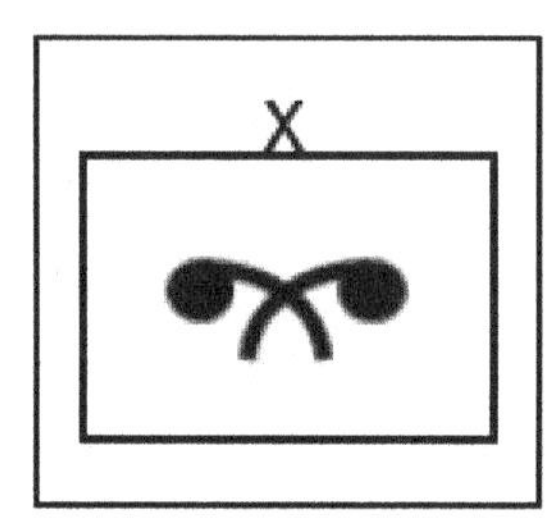

Figure 2-3. CBRN brigade graphic

2-11. The CBRN brigade provides support to corps and echelons above corps. It consists of a headquarters and 2 to 5 battalions. With the advice of his staff and the corps CBRN section, the CBRN brigade commander evaluates and determines the CBRN unit support requirements for the corps. The brigade commander may advise the corps commander on the employment and sustainment of CBRN assets.

2-12. The CBRN brigade offers a range of capabilities to the theater. Units from these brigades can be task-organized to support the COCOM, subordinate joint force commanders, Army force commanders, and functional components faced with CBRN threats or hazards.

2-13. The brigade headquarters provides logistical support and facilitates the coordination of support for subordinate units. This allows the supported unit staff to concentrate on CBRN operations and execution. CBRN brigades lack the capability to provide logistical support to assigned units. CBRN brigades coordinate logistical support and facilities for subordinate units, which allows the staff to focus on CBRN operations.

OFFENSIVE OPERATIONS

2-14. In offensive operations, the CBRN brigade that is task-organized for the mission provides CBRN assets (reconnaissance, hazard response, CBRNE response teams) to the security force. Because the security force operates well forward of the corps main body, these CBRN assets should be attached to the unit that operates as the security force. As the forward line of troops advances, the CBRN brigade allocates units to accomplish the security and stability tasks required to maintain freedom of action in the close area and support to the continuous consolidation of gains. The remainder of the brigade should be task-organized based on METT-TC. The command and support relationships of these task-organized CBRN units are determined based on the ability of the brigade headquarters to coordinate support and provide command and control. Capabilities of the CBRN brigade include—

- Executing or transferring mission command of units assigned to the CBRN brigade and up to five subordinate chemical battalions.
- Providing intelligence support for CBRN operations.
- Planning and coordinating with the division, corps, or theater Army area to which they are assigned or attached.
- Assessing CBRN unit capabilities, utilization, and impacts on plans and operations to the division, corps, and theater Army area.
- Allocating units and resources in support of CBRN reconnaissance, decontamination, and biological detection.

- Operating a tactical command post and a main command post.
- Coordinating sustainment for CBRN operations.

DEFENSIVE OPERATIONS

2-15. The CBRN brigade plans and allocates CBRN units based on the corps mission and the operational and mission variables. The focus should be three-fold—protect the reserve or striking force to support a decisive counterattack, provide support to units in the main battle area, and retain a flexible and responsive CBRN force in the corps support and consolidation areas. Due to an increased risk of CBRN attack, units in the reserve or striking force require support in the form of CBRN reconnaissance to provide early warning to commanders and decontamination support to rapidly reconstitute their combat power. Organic CBRN assets and decontamination capabilities are used to support units in the close area, with enhancement from CBRN units as required. CBRN route reconnaissance is conducted along ground lines of communication from the support area to the forward line of troops to prevent the isolation of forces or early culmination of defensive operations as a result of CBRN attacks.

2-16. CBRN units are allocated from the brigade to assess, protect, and mitigate based on the CBRN threat to the corps support and consolidation areas. CBRN units are in general support. However, their workload is designated through area or task assignments. Requests for immediate CBRN support flow through the corps command post to the CBRN brigade. The CBRN brigade tactical operations center then directs the mission or tasking to the appropriate CBRN unit. Routine CBRN support requests go to the corps CBRN section. The corps CBRN section analyzes the requirement. The corps assistant chief of staff, operations (G-3) assigns a priority to the requirement. The prioritized requirement is passed to the CBRN brigade tactical operations center. The CBRN brigade staff determines the allocation of support necessary to provide the support and issues instructions to the appropriate subordinate unit. Subordinate units are task-organized according to the military decisionmaking process.

BATTALION

2-17. A CBRN battalion supports a division or echelons above division. It consists of a headquarters and two to five CBRN companies. The battalions exist within all three components and provide command and control for companies that have a wide range of capabilities. The CBRN battalion depends on appropriate elements within the theater for religious, legal, force health protection, finance, and personnel and administrative services. See figure 2-4.

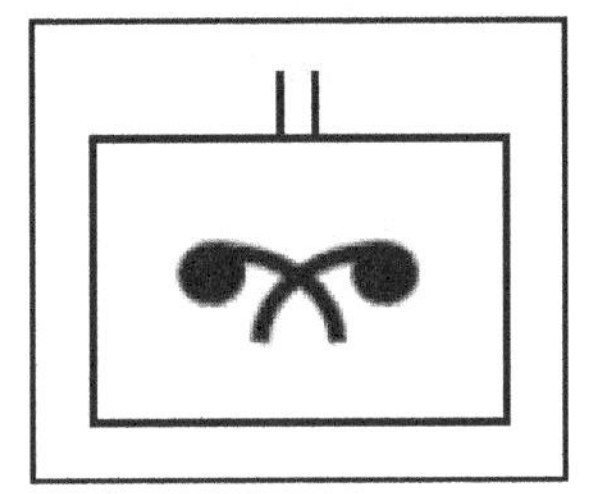

Figure 2-4. CBRN battalion graphic

2-18. The roles, functions, and responsibilities of the CBRN battalion are directly affected by the command and support relationship established with the supported unit. The division commander may place attached CBRN companies under the control of the attached CBRN battalion.

2-19. When a CBRN battalion supports a corps instead of a CBRN brigade, the corps CBRN staff assumes a greater role in planning and coordinating CBRN unit functions, to include the coordination of logistical support to the CBRN units. The battalion advises the supported commander on the employment of the battalion assets.

OFFENSIVE OPERATIONS

2-20. The organization and positioning of the CBRN battalion is based on MDMP and the ability of the CBRN brigade or supported division to provide command and control and logistical support. A CBRN battalion is typically employed in support of a division or higher echelon. A direct support role may be the best to provide the supported unit with rapid and flexible support. A CBRN battalion provides command and control for two to six CBRN companies. Capabilities of the CBRN battalion include—

- Executing or transitioning mission command of personnel assigned to the CBRN battalion headquarters and up to five CBRN companies.
- Providing intelligence support for CBRN operations.
- Planning and coordinating with the brigade, division, corps, or theater Army area to which they are assigned or attached.
- Assessing chemical unit capabilities, utilization, and impacts on plans and operations to the brigade, division, corps, and theater Army area.
- Operating a battalion main command post.
- Coordinating sustainment for CBRN operations.
- Ensuring petroleum, oils, and lubricant distribution and special supply distribution for assigned companies.

DEFENSIVE OPERATIONS

2-21. CBRN battalions are normally positioned based on METT-TC. A CBRN battalion is typically employed in support of a division or higher echelon. Unless there are unusual circumstances, the security force does not require a CBRN battalion headquarters. CBRN battalions should be allocated to support the divisions that conduct the corps main and supporting efforts. At least one battalion headquarters should be allocated to control CBRN assets in the corps consolidation area. The command and support relationships of the battalions that support the divisions are based on providing the supported commander with sufficient versatility to prepare his defense.

2-22. The CBRN battalion provides assets in support of the division consolidation area. The battalion staff must analyze the assigned missions/tasks and develop a support plan that defeats CBRN usage from enemy-protracted resistance and maintains strategic and operational reach to forces in close and deep areas. The battalion may place subordinate CBRN units in general support and provide mission or task assignments or place subordinate elements in a direct support role. The AOR, the threat, and the ability to communicate dictate which technique to use. The battalion may task-organize subordinate companies to form CBRN teams, depending on METT-TC.

2-23. CBRN battalions that support divisions must plan and execute CBRN support operations in the division support area. Units that require immediate CBRN support request it through their higher headquarters, usually one of the divisions major subordinate commands. The CBRN section analyzes and coordinates with G-3 plans to determine future requirements. Based on recommendations by the CBRN staff, the division commander issues guidance. The battalion commander executes the commander's guidance, and the division chemical section provides staff supervision.

COMPANY

2-24. The CBRN company consists of a headquarters and platoon capabilities as described in ATP 3-11.36. The four basic company structures are hazard response, area support, CBRNE, and biosurveillance. See figure 2-5.

2-25. Based on the enemy threat of using ground-contaminating CBRN agents, the CBRN reconnaissance platoon focuses on maintaining freedom of maneuver behind units in close combat. One section should be designated to ensure that the main and secondary routes to and through the area are clear of contamination for each unit. The selection of decontamination sites is coordinated with the supported units and higher headquarters to preclude using a key piece of terrain.

Note. See ATP 3-11.36 for more information on CBRN and CBRNE companies.

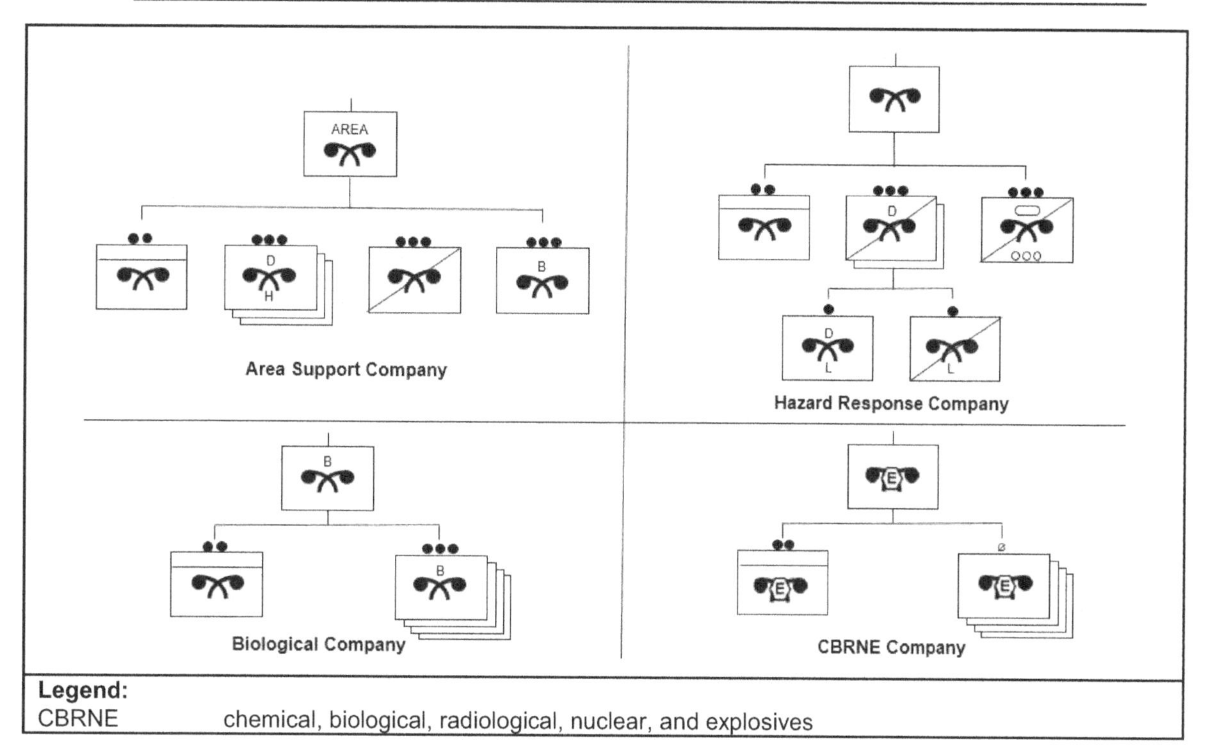

Legend:
CBRNE chemical, biological, radiological, nuclear, and explosives

Figure 2-5. CBRN companies

HAZARD RESPONSE

2-26. A hazard response company provides CBRN reconnaissance, surveillance, and decontamination in support of brigades. A hazard response company can remain task-organized under a CBRN battalion or brigade to support division or corps missions. It has the ability to locate, identify, and mark routes and boundaries around contaminated areas; operate protected within a hazard area; assess hazards in support of site exploitation; and provide field identification of known agents and materials. Their subordinate platoons conduct the following tasks:

- Conduct CBRN reconnaissance and surveillance.
- Conduct operational and thorough decontamination support.
- Conduct site assessment and characterization.

2-27. The hazard response company is aligned to support the BCT and its organic units. The company provides the BCT planners an accurate estimate of their capabilities to support the mission. Although it has decontamination capabilities, it doesn't have the capacity for operations that require decontamination of large numbers of equipment without additional resources. Units need to understand and plan for these limitations.

AREA SUPPORT

2-28. An area support company provides CBRN reconnaissance, surveillance, and decontamination support in theater and corps support areas and consolidation areas. It has the ability to locate, identify, and mark routes and boundaries around contaminated areas. An area support company has a biological surveillance and monitoring capability and the means of providing support to terrain and fixed-site decontamination. Their subordinate platoons conduct the following tactical tasks:

- Conduct CBRN reconnaissance and surveillance.
- Conduct operational and thorough decontamination support.
- Conduct terrain, fixed-site, and aircraft decontamination.

2-29. The area support company is aligned to support the corps. The throughput of the decontamination platoon is greater for the area support company than the hazard response company because of the equipment capabilities.

BIOLOGICAL DETECTION

2-30. A biological CBRN company provides biological surveillance and monitoring of key areas and critical nodes. It can monitor, sample, detect, identify, and report biological agents.

CBRNE

2-31. A CBRNE company consists of specialized teams, including EOD, to provide the advice, assessment, sampling, detection, verification, render-safe, packaging, and escort of CBRN and explosive devices or hazards. When provided security from maneuver elements, CBRNE companies enable combined arms CWMD operations and sensitive site exploitation that feeds intelligence and targeting processes. Their subordinate platoons conduct the following tactical tasks:

- Conduct CBRNE sensitive site exploitation.
- Conduct WMD defeat, disablement, and/or disposal.
- Mitigate CBRNE hazards.
- Escort CBRN samples.
- Escort devices.

2-32. CBRN companies have varied capabilities that exist in their assigned platoons which can be further task-organized. The priority of support is to the lead maneuver units for reconnaissance and surveillance. The offensive tasks (movement to contact, attack, exploitation, and pursuit) require CBRN companies that are task-organized to support through locating and marking contaminated areas and providing assistance in decontamination to allow continued operations while limiting the spread of contamination. Decontamination elements should be prepared to augment maneuver operational decontamination efforts to maintain momentum.

UNITED STATES ARMY SPECIAL OPERATIONS FORCES

2-33. The United States Army Special Operations Command mans, trains, equips, educates, organizes, sustains, and supports forces to conduct special operations across the full ROMO and spectrum of conflict in support of joint force commanders and interagency partners to meet theater and national objectives. Through an indigenous approach, Army Special Operations Forces (ARSOF) interact with other governmental organizations and foreign militaries regularly to shape the environment for Army operations. Operations to shape provide opportunities for conventional forces and SOF interdependence, integration, interoperability, and information sharing. ARSOF is postured to assist in detecting and defeating proliferators and their networks through the ARSOF critical capabilities and during ARSOF core activities. See figure 2-6.

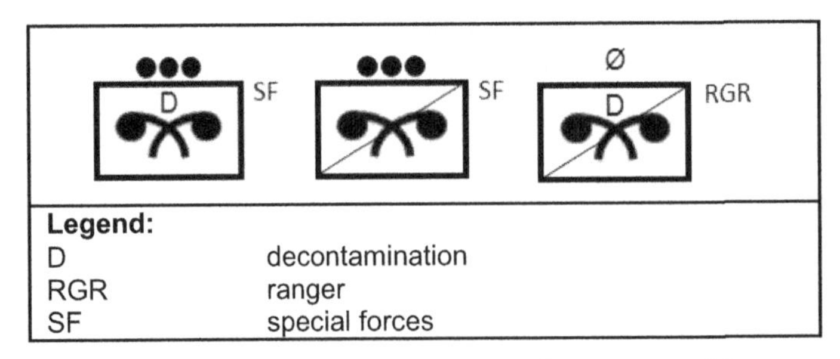

Legend:

D	decontamination
RGR	ranger
SF	special forces

Figure 2-6. Special operations forces graphic

2-34. If adequately integrated into CWMD missions and large-scale combat operations, ARSOF provides capabilities that support CWMD operations. Special operations CWMD operations include WMD interdiction. WMD interdiction is aimed at the early defeat of an adversary WMD program before it matures and focuses primarily on moving targets. Interdiction operations track, intercept, search, divert, seize, or otherwise stop the transit of WMD, its delivery systems, or related materials (including dual use, technologies, and expertise). CWMD operations are actions to disrupt, neutralize, or destroy a WMD threat before it can be used, or to deter the subsequent use of such weapons.

2-35. ARSOF operations and dependencies on CBRN forces or operations must be considered early in joint and Army planning processes. ARSOF does not have the logistics support to operate long-term in a CBRN environment and must rely on the Army to provide some CBRNE capabilities, such as thorough decontamination and individual protective equipment/personal protective equipment resupply.

Note. See ADRP 3-05 for detailed information on ARSOF operational tenets and interdependence. See ATP 3-05.11 and ATP 3-11.24 for more information about special operations in CBRN environments.

CBRN STAFF

2-36. CBRN staffs or elements are the commander's subject matter experts (SMEs) in all levels of command. CBRN staff sections exist at the Army Service component command, corps, division, brigades, special forces group, and battalion levels. The CBRN staff section or element performs the same functions at most echelons of command. In addition to common staff responsibilities, CBRN staff-specific responsibilities fall into the four basic activities of the operations process: planning, preparation, execution, and continuous assessment.

2-37. CBRN staff elements perform many of the same basic duties at most echelons of command as described in ATP 3-11.36. They provide the necessary expertise in understanding CBRN conditions to provide advice to the commander on the necessary information collection, CBRN defense, and response actions to take while conducting offensive tasks. They also integrate information collected from all tasks to gain a better understanding of the CBRN environment and its potential impact on operations.

2-38. The CBRN staff integrates into IPB, bringing their area of expertise to the process. IPB is integral to targeting, risk management, and the information collection process. CBRN vulnerability assessments are the CBRN staff contributions to the IPB process. They bring an understanding of how enemy CBRN COAs will affect the OE and the positioning of CBRN assets. In identifying intelligence gaps during the IPB process, information requirements are established.

2-39. The CBRN staff helps define the OE by identifying the effects of operational and mission variables on operations. Staff planners often utilize PMESII-PT as an analytical starting point to assess the OE, gain a better understanding of a geographic area, and subsequently apply that understanding to achieve objectives.

2-40. Understanding the OE and the threat leads to the information collection strategy and targeting of specific enemy CBRN capabilities. The CBRN staff describes the effects of the identified variables, threat evaluations, and enemy COAs. Terrain and weather also have a profound effect on any CBRN effects that might occur. Civil considerations reflect the impact that industrial facilities, storage sites, and large populations have on CBRN assessments. The assistant chief of staff, civil affairs operations (G-9) and associated civil affairs forces can provide the most up-to-date information on the civil component in the OE.

2-41. The CBRN staff at the corps and division level are primarily involved in the orders process and assist in the development of annex C (Operations) and annex E (Protection). The protection annex includes protection considerations for CBRN hazards, is the result of initial threat and vulnerability assessments, and provides a guide to subordinate units on CBRN protection guidance.

2-42. The CBRN staff functions include the following items and are further described in ATP 3-11.36 and FM 6-0:

- Conduct CBRN threat and vulnerability assessments.
- Recommend courses of action to minimize friendly and civilian vulnerability.
- Coordinate across the entire staff while assessing the effects of enemy CBRN-related attacks on current and future operations.
- Coordinate Army health system support requirements for CBRN operations with the surgeon.

- Establish personnel recovery coordination measures for CBRN environments.
- Advise the commander on the employment of CBRN assets and capabilities.
- Manage the chemical, biological, radiological, and nuclear warning and reporting system (CBRNWRS).
- Model CBRN hazards to assist in gaining hazard awareness and understanding of a CBRN event.
- Advise the commander on the effects of CBRN hazards.
- Assess weather and terrain data to determine the environmental effects on potential CBRN hazards.
- Assist subordinate CBRN staffs and units.
- Incorporate CBRN battalion or brigade staff into MDMP, rehearsals, and commander's updates.
- Integrate into working groups as designated (protection, information collection, and targeting).
- Manage chemical defense equipment.
- Assist units with acquiring medical CBRN defense materiel.

2-43. CBRN staffs at brigade, battalion, and SOF units focus on support to the commander's scheme of maneuver to support the commander's intent and concept of operation. Brigade, special forces group, and battalion CBRN officers and NCOs integrate CBRN assets into mission planning and ensure that organic CBRN assets are trained and synchronized to support maneuver units. CBRN brigade coordination elements and battalion staffs should integrate into MDMP, rehearsals, and commander's updates with corps and division CBRN staff. Additionally, they train and provide guidance to CBRN specialists at the company, battery, or troop/detachment level.

2-44. CBRN staffs focus on assisting and advising the commander, contribute to timely decision making by the commander, and supervise the execution of decisions. They advise and make recommendations based on their running estimates, enabling the commander to assess and manage risk and to consider which vulnerabilities to accept or mitigate. CBRN staffs build and maintain running estimates to track and record information focused on their specific area of expertise and how CBRN elements are postured to support future operations. Examples of facts and assumptions that may impact operations from the CBRN perspective include—

- Adversarial CBRN employment and the effects of CBRN hazards.
- Potential weather effects.
- Named areas of interest with suspected CBRN hazards.
- Industrial facilities that could create hazards if impacted by collateral damage.
- Friendly and partner nation CBRN unit status.

2-45. CBRN staffs support the commander in communicating decisions and intentions through plans and orders. The commander's activities of understand, visualize, describe, and direct are supported by the running estimates provided by CBRN staffs and the subject matter expertise provided by CBRN forces at every echelon. See appendix D for an example of a CBRN running staff estimate.

2-46. CBRN staffs at all levels must frequently work together to monitor, improve, and sustain CBRN training, unit readiness, and standard operating procedures. CBRN staffs also plan and coordinate the employment of supporting CBRN units.

THREAT ASSESSMENTS

2-47. IPB allows the commander to visualize the enemy situation and discern the enemy commander's probable intent. Identifying threat capabilities, strengths, and weaknesses and assessing intent are critical to providing commanders early and accurate warnings of threat actions, which is necessary to ensure operational success. Recognizing adversary capabilities, setting the conditions to win future conflicts, and supporting U.S. and partner nation capabilities to protect themselves rely on accurate assessments. The CBRN staff must assist in the overall CBRN threat template and identify vulnerabilities to the mission and forces.

2-48. CBRN staffs maintain running estimates, continuously updating threat assessments as information is received. These assessments help to understand the current OE and to counter CBRN threats. The threat assessment process is continuous and integrated into the operations process. It informs the commander's CCIR and determines where information collection capabilities should be allocated. Assessments help the commander develop, adapt, and refine plans.

2-49. CBRN staffs provide expertise to the G-2/S-2 to answer questions about CBRN hazards that limit the commander's options. CBRN threat assessments help estimate the probability of the use of WMD, which is input for the vulnerability analysis process. The CBRN threat assessment helps the commander understand the presence of CBRN threats and hazards (such as TIMs). The threat assessment provides a framework for mission planning. The threat assessment process is continuous and directly tied to the commander's decisions throughout the conduct of operations.

Note. See ATP 3-11.36 for more information on CBRN assessment activities.

Example Impact of Threat Assessment

CBRN staff at the division level, integrated into the planning process, analyze the likelihood that the enemy will employ CBRN threats. They prepare a threat assessment that a potential hazard exists based on a history of the enemy employing CBRN weapons at wet gap-crossing sites and breaching operations to defend key terrain. The staff develops courses of action to provide an early warning of a CBRN attack. An information collection plan is developed to acquire more information about potential CBRN attack locations. CBRN reconnaissance assets are integrated into the collection plan and tasked to answer CCIR.

The organic heavy CBRN reconnaissance platoon assigned to a brigade armored reconnaissance squadron is assigned the named area of interest covering a wet gap-crossing site. The nuclear, biological, and chemical reconnaissance vehicle platoon, nested with a cavalry troop, establishes a listing post/observation post within the range of its sensors. During operations, the nuclear, biological, and chemical reconnaissance vehicle platoon submits a CBRN 1 report of a suspected CBRN attack through indirect fires at the NAI.

OFFENSIVE OPERATIONS

2-50. CBRN staff considerations for the offensive operations include—

- Focusing CBRN defense to provide the commander flexibility and facilitate synchronization.
- Planning operational decontamination, as necessary.
- Planning thorough decontamination after the mission is complete.
- Selecting decontamination sites throughout the zone.
- Advising on the lowest possible protective posture based on mission-oriented protective posture (MOPP) level analysis.
- Identifying known or suspected areas of contamination.
- Focusing CBRN reconnaissance assets to retain the freedom of maneuver.
- Prioritizing CBRN assets to lead maneuver forces.
- Balancing vulnerability of the force against the need for mass and speed.
- Identifying possible contaminated areas and possible by-pass routes.
- Considering the possibility of increasing enemy CBRN attacks as the attack progresses.
- Identifying NAIs as part of the information collection plan.
- Integrating CBRN assets into the reconnaissance and surveillance scheme.

> **CBRN Support to Intelligence**
>
> CBRN forces support the intelligence warfighting function at every echelon. At corps and division levels, CBRN staffs contribute to the intelligence preparation of the battlefield process. These CBRN staff officers lend their CBRN subject matter expertise to developing estimates on the threat to add COA development. The CBRN staffs assist with allocating the limited CBRN assets available to maximize the effectiveness of CBRN reconnaissance, surveillance, and decontamination.
>
> At the battalion level, CBRN staff continue to be a part of MDMP and have a more direct involvement in the liaison between CBRN assets and maneuver force mission planning.
>
> At the tactical level, CBRN forces act on the plans for information collection planned by the staffs. CBRN reconnaissance efforts are directed toward answering CCIRs. Assessments of WMD sites feed information back into the intelligence process. Staffs take the results of reconnaissance efforts to update estimates and revise information collection efforts.

Defensive Operations

2-51. Commanders and staffs deliberately plan and integrate protection capabilities to protect the force, preserve combat power, reduce risk, and mitigate vulnerabilities. Protecting the force supports the ability to regain the initiative.

2-52. All three defensive tasks (area defense, mobile defense, and retrograde) use terrain as force multipliers. Terrain influences the tempo of enemy attacks and provides the defender with cover and concealment. The CBRN staff must understand and communicate the impacts terrain has for protection from and influence of CBRN conditions.

2-53. Staff considerations include—

- Focusing CBRN defense to provide the commander versatility and synchronization.
- Operating in the lowest possible MOPP level during the preparation phase, then considering a higher MOPP for the actual battle.
- Selecting decontamination sites throughout the support area to support the defensive scheme.
- Conducting operational decontamination as necessary for survivability.
- Focusing CBRN reconnaissance assets on repositioning and withdrawal routes, passage points, and passage lanes.
- Identifying alternate routes if passage routes become contaminated.
- Designating passage points and lanes for the movement of contaminated elements.
- Balancing vulnerability of the force against the need for mass, agility, and depth.
- Understanding the enemy order of battle and doctrinal templates as they relate to CBRN use.
- Executing thorough decontamination as necessary after the battle.

Organic CBRN Teams and Platoons

2-54. CBRN capabilities exist down to the unit level. Reconnaissance and surveillance platoons exist in BCTs, and unit CBRN teams are formed at companies.

Unit CBRN Teams

2-55. Duties of unit CBRN teams are not necessarily performed by Soldiers with military occupational specialty 74D. At company, battery, and troop levels, trained CBRN specialists provide advice and support to operations in CBRN environments to the commander. These Soldiers integrate CBRN conditions into unit mission essential tasks to mitigate impacts on unit operations. The unit level CBRN specialist maintains company equipment, training, and programs for CBRN operations and can form the basis for training unit teams to conduct CBRN detection, survey for contamination during AA site selection, and conduct decontamination. Unit CBRN teams are trained and ready to send CBRN 1, spot reports, or other reports required by unit standard operating procedures.

Note. See FM 6-99 for CBRN and spot report formats.

2-56. To facilitate movement and conduct maneuver, unit CBRN specialists can assist in developing unit level CBRN defense plans. Assets within the organization can provide operational decontamination, augment detailed equipment decontamination, and conduct detailed troop decontamination. Additional support for thorough decontamination can be requested from CBRN units.

Reconnaissance and Surveillance Platoons

2-57. The Army maintains two variants of the CBRN R&S platoons in the modular force design. The primary differences are the way in which they conduct missions and the vehicles they use to conduct those missions. The CBRN R&S platoon (light) primarily conducts R&S by using the dismounted reconnaissance sets, kits, and outfits. They are capable of conducting missions in urban environments where maneuver is confined. The CBRN R&S platoon (heavy) conducts mounted reconnaissance using a CBRN reconnaissance vehicle. This more mobile platform allows heavy platoons to cover more terrain in support of route and area reconnaissance. See figure 2-7.

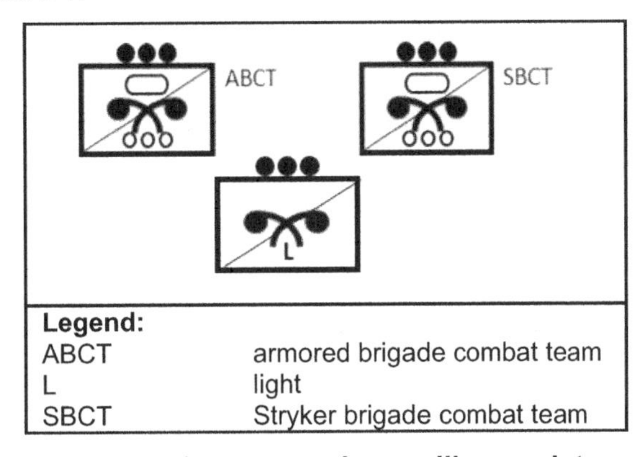

Figure 2-7. Reconnaissance and surveillance platoon graphic

OTHER ORGANIZATIONS

2-58. There are numerous CBRN organizations that have unique and specialized capabilities which exist across the force. CBRN reconnaissance and decontamination teams exist with special operations forces. Laboratory capabilities and nuclear disablement teams are a part of the CBRNE command (see ATP 3-37.11). The following list and paragraphs 2-54 through 2-63 provide an example of some of the organizations with ties to CBRN operations:

- Maneuver enhancement brigades (MEBs). See FM 3-81.
- Other Service CBRN elements.
  - Air Force Radiological Assessment Team.
  - Marine Corps Chemical, Biological, Incident Response Force.
- Interagency partners.
  - Mobile uranium facilities.
  - Mobile plutonium facilities within the Department of Energy.
- Reachback and technical support.

- The Edgewood Chemical Biological Center provides reachback technical support for chemical and biological hazard information.
- The United States Army Medical Research and Materiel Command (USAMRMC) operates labs that provide operational, strategic, and reachback capabilities for chemical and biological (CB) threats.
- The United States Army Research Institute of Chemical Defense (USAMRICD) focuses on chemical warfare agents and training on the medical management of chemical casualties.
- The United States Army Medical Research Institute of Infectious Disease (USAMRIID) focuses on biowarfare agents and infectious pathogens.
- The CBRN Information Resource Center provides current and authoritative information on the DOD Chemical and Biological Defense program through the Information Resource Center and the Joint Acquisition CBRN Knowledge System.
- The Defense Threat Reduction Agency (DTRA) provides expertise via reachback when the capabilities required exceed the capabilities of the organic CBRN staff. DTRA supports combatant commands, U.S. military and partnered nations, the federal government, academic institutions, first responders, and the international community.
- The National Ground Intelligence Center (NGIC) is the Defense Department's primary producer of ground forces intelligence. NGIC has highly skilled specialists and other technical specialists who evaluate the capabilities and performance data on virtually every weapons system used by a foreign ground force, including chemical and biological weapons and future weapons concepts.
- The United States Army Nuclear and Countering Weapons of Mass Destruction Agency (USANCA) is part of the Army, G-3/5/7 and is a field-operating agency. USANCA provides U.S. nuclear employment guidance to ground forces and CWMD planning expertise for the implementation of Army CWMD strategy and policy at the corps level and above.

Note. See ATP 3-05.11, ATP 3-11.24, and ATP 3-11.36 for more information on other CBRN forces, other Service CBRN forces, and other reachback agencies.

SECTION II—TRAINING

2-59. Commanders and their units, down to the lowest private at the tactical level, must be capable of surviving CBRN attacks and continuing operations in hazardous conditions. More information about the basic proficiency skills for individuals and teams can be found in appendix E.

2-60. Training is the interaction among three separate but overlapping training domains (operational, institutional, and self-development) synchronized to achieve the goal of trained Soldiers, leaders, and ready units. The content of this publication focuses on the operational training domain and building CBRN readiness.

THE OPERATIONAL TRAINING DOMAIN

2-61. The operational domain encompasses training activities that unit leaders schedule and individual units and organizations undertake. Unit leaders are responsible for the proficiency of their subordinates, subordinate leaders, teams/crews, and the unit as a whole to assess, protect, and mitigate CBRN hazards within the scope of their missions. Subordinate leaders assist commanders in achieving training readiness proficiency goals by ensuring that training is conducted to standard, with CBRN as a complex mission variable, in support of the unit mission essential task list. The mission essential task list-based strategies are known as the Combined Arms Training Strategy database and synchronized with missions to build and sustain unit readiness.

2-62. Commanders use their standardized mission essential task list, with CBRN conditions incorporated into the training, to achieve greater proficiency in a contested CBRN environment. Unit training plans use a crawl-walk-run approach that progressively and systematically builds on successful task performance before progressing to more complex tasks. This approach enables a logical succession of enabling skills and knowledge—from basic to advanced tasks and conditions—for individual, crew, and collective training. Training must be challenging, integrating increasing levels of complexity, stress, and duration against conventional, hybrid, and asymmetric threats in the OE to return units to mastery at CBRN operations.

2-63. Realistic training with limited resources demands that commanders focus their unit training efforts to maximize repetitions under varying conditions to build proficiency. Readiness is built on the ability to be proficient regardless of the conditions, to include CBRN environmental conditions. Operating during CBRN contamination should be a training condition, not simply a task. This means that, while commanders focus their training efforts on primary battle tasks, CBRN training is integrated as a condition, not as a separate event. Training should incorporate hazards, including improvised explosive devices, CBRN effects, the impacted population, cultures and languages, key leader's decisions, joint and domestic partners, and special operations forces. It should also portray the impact that decisions will have on mission success.

Note. See FM 7-0 for more information.

Vignette

This training requirement includes the requirement to conduct a live-fire while wearing protective equipment. Rather than requiring the unit to put on a mask and gloves to fire, the unit incorporates a scenario of reacting to a chemical attack into squad and platoon qualification tables. After the unit achieves MOPP 4 and uses the CBRN warning and reporting system to report the incident, the unit must continue the mission. The unit completes the requirement for firing weapons while wearing protective equipment. Upon completion, the unit conducts unmasking procedures (see STP 21-24-SMCT) before giving the ALL CLEAR.

TRAINING IDEOLOGY

2-64. The anticipated OE will require Army forces to operate as combined arms teams, in joint operations, and with multinational forces. Similarly, CBRN units should frequently integrate training with other combined arms units, building habitual relationships that endure during large-scale combat operations. Combined training exercises require organizing and synchronizing with increasing complexity to stress the rigors of operating in a CBRN environment. Combined arms training allows planning to address the additional considerations of CBRN hazards from best practices, lessons learned, and task organizations to align with mission requirements.

2-65. Corps and division CBRN cells advise and assist brigade CBRN staffs with training, providing best practices, and assisting in planning and resourcing. The division CBRN cell also provides oversight and support to CBRN troop schools, providing reachback to the CBRN school.

2-66. CBRN staffs at brigades and battalions (along with company, battery, and troop CBRN specialists) provide commanders with CBRN training oversight, advisement of CBRN readiness, and training recommendations. Battalion and brigade CBRN staffs provide frequent staff assistance visits to ensure that unit CBRN rooms are maintained to organizational inspection program standards. To increase interoperability between units in CBRN reconnaissance and decontamination, CBRN staffs and CBRN unit commanders should coordinate with one another, enhancing training.

2-67. Company commanders and first sergeants leverage the company, battery, or troop CBRN specialist to maintain CBRN equipment readiness in unit CBRN rooms and to incorporate CBRN training into training plans. Platoon leaders and platoon sergeants should conduct CBRN readiness inspections and incorporate CBRN as a variable during collective platoon training in walk and run phases. Chlorobenzylidenemalononitrile (CS) chambers to build mask confidence are also opportunities for Soldiers to train on CBRN-focused warrior tasks and battle drills. Individual CBRN tasks that ensure the survivability of Soldiers are trained by unit NCOs in teams and squads.

2-68. Training programs differ to some extent between headquarters, services, and nations; however, training within North Atlantic Treaty Organization (NATO) programs and partner nations is encouraged so that a similar model is used each time, ensuring a common approach and mutual support for achieving interoperability. In multinational training, knowledge of the CBRN defense capability and practices of other nations is required.

Chapter 3
Supporting Decisive Action

This chapter provides information on CBRN environments and the employment of CBRN forces in support of decisive action. It is divided into four sections: CBRN contribution to offense, defense, stability, and DSCA.

OVERVIEW OF CBRN SUPPORT TO DECISIVE ACTION

3-1. The heart of the Army's operational concept is decisive action. *Decisive action* is the continuous, simultaneous execution of offensive, defensive, and stability operations or defense support of civil authorities tasks (ADP 3-0). As a single, unifying idea, decisive action provides direction for an entire operation. The simultaneity of offense, defense, and stability tasks is not absolute at every level. The higher the echelon, the greater the possibility of simultaneous tasks. Operations conducted outside of the United States and its territories simultaneously combine three elements—offense, defense, and stability. Within the United States and its territories, decisive action combines the elements of DSCA and offense and defense to support homeland defense, when required.

3-2. CBRN units operate as part of larger combined arms task forces or teams to enable the freedom of action during decisive action. Due to the mix of capabilities within CBRN units, CBRN planners must ensure that the right units and capabilities are applied to execute specific tactical tasks.

3-3. CBRN units have sufficient fire power to conduct local security. *Local security* is a security task that includes low-level security activities conducted near a unit to prevent surprise by the enemy (ADP 3-90). Local security provides immediate protection to friendly forces and is typically performed by a unit for self-protection, but it may be provided by another unit when the security requirements are greater than the unit security capabilities. Throughout decisive action, CBRN units operate within the supporting range of other units. *Supporting range* is the distance one unit may be geographically separated from a second unit yet remain within the maximum range of the second unit's weapons systems (ADP 3-0). Table 3-1 provides an example of where specific CBRN units support tactical tasks of decisive action.

Table 3-1. Example CBRN unit support to decisive action tasks

Units	***Offense***				***Defense***			***Stability***	***DSCA***
	Movement to contact	Attack	Exploitation	Pursuit	Area defense	Mobile defense	Retrograde		
Biological					X	X	X	X	X
Area support					X	X	X	X	X
Hazard response	X	X	X	X	X	X	X	X	X
CBRNE (CRT)								X	X

Notes.
1. The "Xs" represent the center of gravity for support to the associated tactical task.
2. During offense and defense tasks, CBRNE, biological, and area support companies support the consolidation of gains but are not directly involved in the tactical tasks listed. See table 2-1 for tasks conducted by these units.

Legend:

CBRNE	chemical, biological, radiological, nuclear, and explosives
CRT	CBRNE response team
DSCA	defense support of civil authorities

3-4. In large-scale combat operations against a peer threat, commanders conduct decisive action to seize, retain, and exploit the initiative. This involves the orchestration of many simultaneous unit actions in the most demanding of OEs. CBRN tasks within large-scale combat operations introduce levels of complexity, lethality, ambiguity, and hindrance to military activities not common in other operations. These operations require the execution of multiple tasks synchronized and converged across multiple domains to defeat enemy forces, control terrain, protect populations, and preserve the freedom of action.

3-5. Army forces generally constitute the preponderance of land combat forces organized into corps and divisions that execute decisive action tasks. Commanders at the corps and division levels are directly concerned with those enemy forces and capabilities that can affect their current and future operations. Army forces use mobility, protection, and firepower to strike the enemy from multiple directions, denying the enemy the freedom to maneuver and creating multiple dilemmas that the enemy commander cannot effectively address.

3-6. Thorough CBRN planning in support of decisive action involves conducting assessments of the OE, threats and hazards, vulnerabilities, capabilities, and risks. Through this planning process, decisions are made for weighing the CBRN functions of assess, protect, and mitigate to best support the tactical tasks. The CBRN staff considers the commanders guidance and the entire AO to determine ways to assess CBRN threats and hazards that may exist in the OE, provide protection of friendly forces, and mitigate CBRN incidents to maximize combat power. CBRN brigade and battalion staffs synchronize and integrate with the corps and division CBRN staff to plan for and execute decisive action tasks. Through successful staff integration, corps and division commanders are provided with more accurate enemy CBRN composition and disposition. This helps determine which defeat mechanisms may trigger increased risk for CBRN attack.

3-7. CBRN staffs further integrate with fires to support the targeting of enemy CBRN capabilities in deep areas. CBRN staffs help determine the emplacement of friendly CBRN units to protect critical assets and to recommend tactical dispersion to minimize enemy CBRN attack effects.

3-8. Army forces assist in CWMD operations within the joint operations area and throughout the area of responsibility during large-scale combat operations. According to the joint force commander's concept of the operation, Army forces plan and integrate their efforts with the joint force and host nation for CWMD operations. CBRN forces contribute to the mission success of CWMD with the understanding that special considerations exist concerning CBRN environments and WMD objectives. For example, planning for CBRN site exploitation; contaminated route management; prolonged operations in CBRN environments; and immediate, operational, and thorough decontamination of units.

Note. More about CBRN planning can be found in ATP 3-11.36. See ATP 3-90.40 for more information on combined arms CWMD operations.

3-9. Corps and division headquarters assign purposefully task-organized forces designated consolidation areas to defeat enemy protracted means of resistance as part of large-scale combat operations. Optimally, a corps commander would assign a division to secure its consolidation area, and a division commander would assign a BCT to secure its consolidation area. Units begin consolidation of gains activities after achieving a minimum level of control and when there are no ongoing large-scale combat operations in a specific portion of their AO. The focus is initially on combined arms operations against bypassed enemy forces, defeated remnants, and irregular forces. Area support companies and CBRNE companies support the corps and division consolidation areas by rapidly exploiting abandoned enemy CBRN material, isolating hazardous material that hampers maneuver, and mitigating post-CBRN strike effects on populations and key terrain. The assigned unit commander eventually concentrates on all stability mechanisms based on the mission objectives, endstate, and duration of the assignment for the consolidation area.

3-10. BCTs and subordinate echelons concentrate on performing offensive and defensive tasks and necessary tactical enabling tasks. During large-scale combat operations, they perform only those minimal essential stability tasks necessary to comply with the laws of land warfare. Hazard response companies are the primary unit for CBRN support to multifunctional brigades and BCTs within the close area. Hazard response companies enhance a brigade's ability to assess and mitigate CBRN hazards.

3-11. In CBRN operations, the tasks and activities conducted occur simultaneously in support of different components of decisive action. The balance of decisive action tasks depends on the focus of the mission (see figure 3-1). The following sections discuss the functions and tasks that are the key component of each of the tactical tasks and the echelons affected.

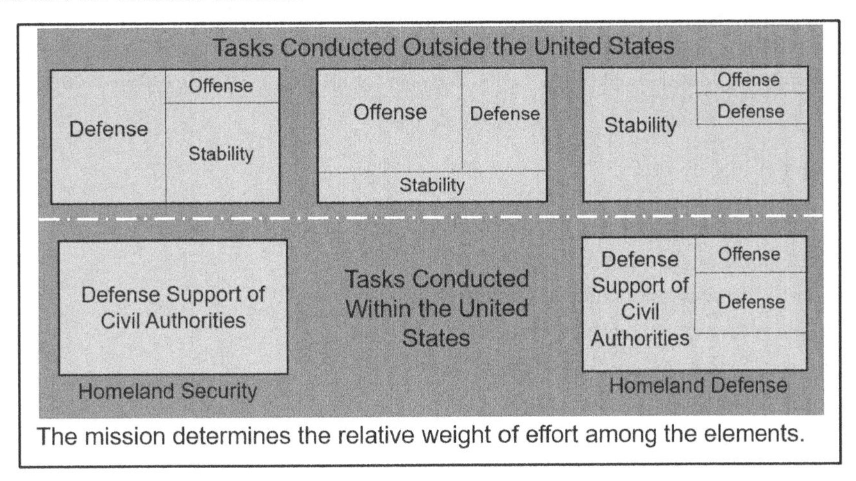

Figure 3-1. Decisive action

3-12. Figure 3-2 provides an example of how CBRN functions are combined to support decisive action. For example, in the offense, tasks that fall into the function of assess (CBRN reconnaissance) weigh more heavily than in a DSCA, where mitigating CBRN incidents is more prominent. Depending on the echelon and the specific activity being conducted, the balance of tasks may be different than those depicted.

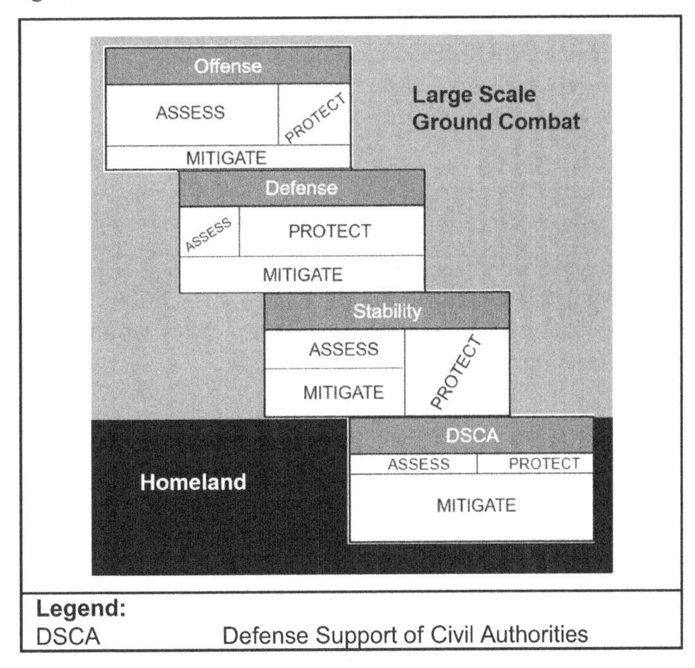

Figure 3-2. Example balance of CBRN functions to decisive action

3-13. Throughout decisive action, CBRN brigades are assigned direct support relationships to corps headquarters, ensuring that subordinate CBRN battalions are task-organized and receive combat service support to operate in support of tactical units. CBRN battalions are direct support to divisions and can task-organize hazard response, area support, biosurveillance, and CBRNE companies as required to provide assets in support of division missions. Hazard response companies are direct support to BCTs and multifunctional support brigades within division close and deep areas.

TACTICAL ENABLING TASKS

3-14. Commanders direct tactical enabling tasks to support the performance of all offensive, defensive, and stability tasks. Tactical enabling tasks are employed by commanders as part of shaping operations. The tactical enabling tasks are reconnaissance, security, troop movement, relief in place, passage of lines, encirclement operations, and mobility and countermobility operations. The CBRN functions (assess, protect, and mitigate) support all tactical enabling tasks; however, they are most utilized in reconnaissance, security, and mobility.

RECONNAISSANCE

3-15. There are seven fundamentals of successful reconnaissance. Commanders—

- Ensure continuous reconnaissance.
- Do not keep reconnaissance assets in reserve.
- Orient on the reconnaissance objective.
- Report information rapidly and accurately.
- Retain the freedom of maneuver.
- Gain and maintain enemy contact.
- Develop the situation rapidly.

3-16. There are five forms of reconnaissance operations: route, zone, area, reconnaissance in force, and special reconnaissance. CBRN forces support to route, area, and zone reconnaissance is described in ATP 3-11.37.

3-17. *Route reconnaissance* is a directed effort to obtain detailed information of a specified route and all terrain from which the enemy could influence movement along that route (ADP 3-90). CBRN route reconnaissance is directed along key ground lines of communication from support and consolidation areas to the forward line of troops. Enemy CBRN attacks can force early culmination and reduce the operational reach of the forces. CBRN route reconnaissance detects, identifies, and marks contamination and pursues an uncontaminated bypass, allowing maneuver forces to continue movement.

3-18. *Zone reconnaissance* is a form of reconnaissance that involves a directed effort to obtain detailed information on all routes, obstacles, terrain, and enemy forces within a zone defined by boundaries (ADP 3-90). CBRN zone reconnaissance supports the CCIRs and designated NAIs.

3-19. *Area reconnaissance* is a form of reconnaissance that focuses on obtaining detailed information about the terrain or enemy activity within a prescribed area (ADP 3-90). CBRN area reconnaissance is executed through the assessment of potential CBRN targets, surveillance of key terrain, and assessment of hazardous material sites that impact maneuver.

SECURITY OPERATION

3-20. Security includes all measures taken by a command from enemy observation, sabotage, annoyance, and surprise. CBRN R&S, in support of a screen security task, attach to cavalry platoons to observe, identify, and report enemy indicators, preparations, or usage of CBRN, enabling friendly forces to retain the freedom of maneuver at tempo. CBRN R&S assets can be attached or tactical control to reconnaissance elements in support of the reconnaissance objective. NAIs are identified during IPB and direct where CBRN reconnaissance should focus sensors. The purpose of CBRN reconnaissance is to confirm or deny the presence of potential or suspected CBRN threats/hazards. CBRN reconnaissance may be augmented by indirect fires and close air support, but otherwise fights only in self-defense. CBRN R&S may augment forces to execute stationary and moving screens, when required. The hazard response company nuclear, biological, and chemical reconnaissance vehicle platoons may be task-organized to support R&S squadrons and troops by section to maximize CBRN assessment in support of the reconnaissance objective. The CBRN R&S platoon detects hazards at a distance, allowing commanders early warning of chemical hazards. See figure 3-3.

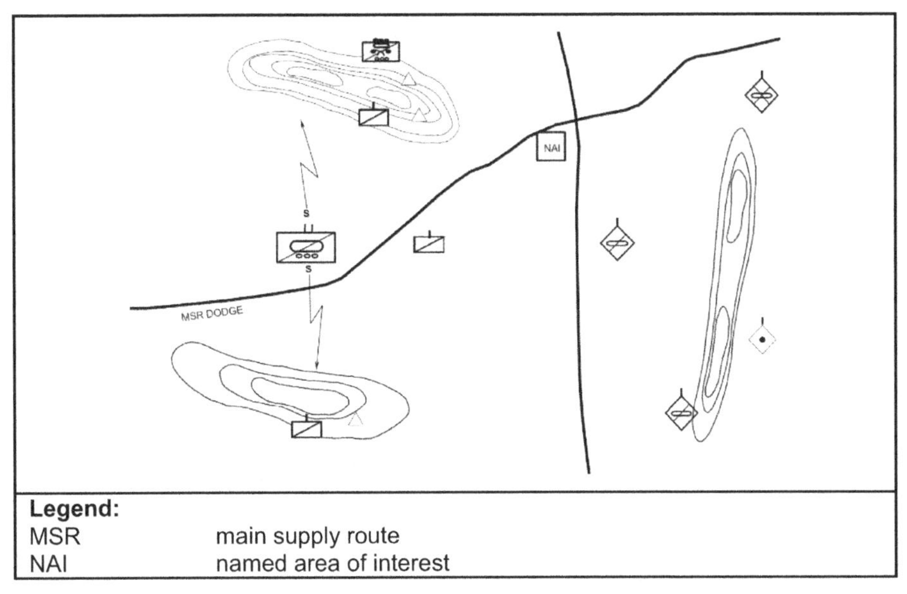

Figure 3-3. Example CBRN reconnaissance in screen

3-21. A guard security task requires a guard force that contains sufficient combat power to defeat, cause the withdrawal of, or fix the lead elements of an enemy ground force before it can engage the main body with direct fire. A guard force engages enemy forces with direct and indirect fires. CBRN reconnaissance assets can be embedded with R&S troops, but they may be placed in a position to provide early warning and bypass of CBRN attacks if enemy forces use CBRN in an attempt to isolate and defeat cavalry squadrons. NAIs for CBRN reconnaissance include canalizing terrain and key lines of communication that CBRN attacks could sever, forcing operational culmination of the security force. See figure 3-4.

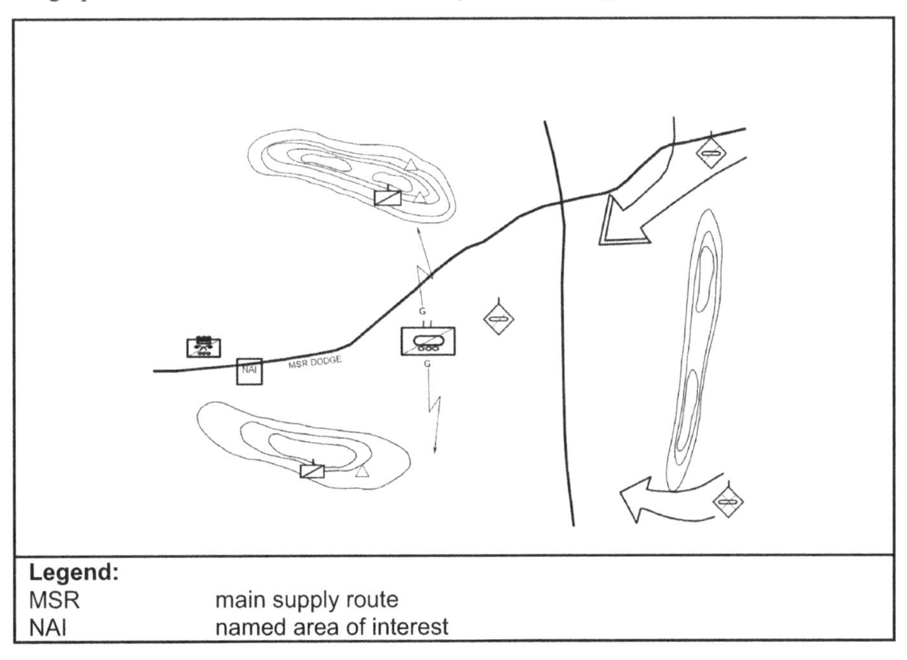

Figure 3-4. Example CBRN reconnaissance in guard

3-22. A cover security task is a self-contained force capable of operating independently of the main body, unlike a screen or guard. CBRN reconnaissance assets have limited protection and a greater likelihood of becoming decisively engaged in a cover. Commanders should make prudent risk decisions to place CBRN reconnaissance assets within a covering force.

MOBILITY

3-23. *Mobility* is a quality or capacity of military forces which permits them to move from place to place while retaining the ability to fulfill their primary mission (JP 3-17). CBRN planners should anticipate the needs of maneuver forces to enhance mobility in CBRN environments. The placement of reconnaissance in and around gap crossings provides rapid assessments of hazards in the event of a CBRN attack. An assessment of the operational risk is central to the planning of gap crossing in a situation compounded by CBRN conditions. Because gap crossing tends to canalize forces and severely restrict maneuver within the crossing area, it is critical to identify and quantify the risk of a CBRN attack on the crossing. Decontamination sites must be incorporated into the planning of gap crossing. The principles of speed (decontaminate only what is necessary), priority (decontaminate the most essential items first), and limited area (decontamination as close to the site of contamination as possible to limit spread) should guide planning. CBRN tasks of decontamination allow for the rapid reconstitution of combat power and prevent the spread of contamination.

SECTION I—CBRN CAPABILITIES IN OFFENSIVE OPERATIONS

OVERVIEW

3-24. During large-scale combat operations, commanders must consider the potential for future enemy actions that might include the use of CBRN. Planners anticipate the potential for enemy capabilities in each of the CBRN hazards to guide the commander's COAs. A planner considers continuing operations in CBRN environments to develop sustainment plans that allow the commander to capitalize on a challenge and turn it into an advantage. Tasks executed during the offense are to exploit the initiative to gain physical and psychological advantage over the enemy or adversary.

3-25. Task organization is the act of designing a force, support staff, or sustainment package of specific size and composition to meet a unique task or mission. There are never enough CBRN units to handle all tasks. CBRN units are task-organized across the battlefield, but they are concentrated with the priority of effort or at the point of need to ensure success. This requires accepting risk elsewhere. CBRN units, teams, and elements must remain flexible to provide the required capability that the mission dictates.

3-26. In operations involving WMD, action or inaction at the tactical level can have profound strategic repercussions. While there has been greatly increased emphasis on CWMD, it is important for commanders and planners to understand that the WMD is not an adversary. It is a capability that an adversary may use to coerce or deter actions or to achieve effects during operations. Thus, CBRN operations should not be considered a special or distinct set of activities or a separate mission area to be used only under certain conditions. Instead, commanders must consider CBRN operations as part of their operational planning for offensive operations. According to the tasks described in table 3-1, CBRN integrates into combined arms teams to execute decisive action tasks.

THREAT OVERVIEW

3-27. The threat overview describes how threat forces organize against U.S. forces in the offense. This allows CBRN forces to understand potential enemy courses of action.

3-28. The peer threat recognizes the defense as the stronger form of military action, particularly when faced with a superior adversary. It sees the offense as the decisive form of military action, however, it is likely that one or more tactical-level subordinate units may execute defensive missions to preserve offensive combat power in other areas. During defensive tasks, the enemy typically attempts to slow and disrupt friendly forces with obstacles, prepared positions, and favorable terrain so that they can be destroyed with massed fires. The enemy is likely to conduct a mobile defense when capable, using a series of subsequent battle positions to achieve depth in conjunction with short, violent counterattacks and fires. The enemy can be expected to employ significant electronic warfare (EW), intelligence, surveillance, reconnaissance, and information-related capabilities as part of this defensive effort. Several potential enemies have a chemical weapons capability, and some could employ tactical nuclear weapons.

3-29. The enemy typically organizes for operations with a forward disruption force followed by a main defense force with at least one tactical reserve. A linear battlefield starts with a disruption zone in front of the battle zone with a support zone in the rear. Disruption forces occupy the disruption zone to slow and degrade an adversary before it can make contact with the main defense force. The battle zone is where the main defense force establishes defensive arrays with kill zones or complex battle positions to further disrupt, fix, turn, block, or destroy its opponent. Inside the battle zone, reserve forces are postured in battle positions or AAs and are usually strong enough to defeat an adversary's exploitation force. The support zone includes the enemy's main command posts, long-range fires, air defense, and sustainment assets.

3-30. The enemy prefers mobile defense in situations for which it is not overmatched. A disruption force and/or a main defense force (or part of it) can perform defensive maneuver. In either case, the force must divide its combat power into two smaller forces: a contact force and a shielding force. The contact force occupies the defensive array in an area that is in current or imminent contact with its opponent. The shielding force occupies the portion of the defensive array that permits the contact force to reposition to a subsequent array. The disruption force initiates the attack on its adversary by targeting and destroying critical combat systems. The mission of the main defense force is to complete the defeat of its opponent by a more robust counterattack force.

3-31. An area defense is used by the enemy when it is overmatched by its adversary or when it must deny key terrain to its adversary. The disruption force continues to harm its opponent with observed indirect fires from the support zone without causing significant exposure to its own forward forces. The disruption force accomplishes this by using a series of concealed observation posts that are spread in-depth across the disruption zone. The main defense force occupies complex battle positions within the battle zone. These battle positions are designed to prevent opponents from employing concentrated direct fires without costly maneuver in kill zones tied to terrain and obstacles.

3-32. The enemy uses CBRN threats to support its defense operations when its opponent is preparing for attack and during main engagements. The threat's use of nonpersistent or persistent chemical agents, rather than biological or nuclear, is more likely because enemy forces can anticipate and control the effects. Typical target options for chemical agents are AAs, along favorable axes of advance, and routes that support the employment of a reserve.

3-33. The threat usually retains authority for the employment of biological agents at the national level because of political ramifications and its ability to control a subsequent epidemic. Probable targets for biological warfare pathogen attacks are airfields, logistics facilities, population centers, and command centers deep within an adversary's rear area to prevent the release from affecting its own forces.

3-34. The threat may develop and employ radiological weapons. The effects of these weapons are achieved by using toxic radioactive materials against desired targets. Like biological warfare, radiological weapons are considered weapons of intimidation and terror. They can achieve area denial if delivered in high concentrations; otherwise, they have a disruptive effect and are employed at locations that support the synchronization of other threat assets.

3-35. If the threat is capable of delivering nuclear weapons, it retains a release authority at the national level. The primary use of nuclear weapons in the defense is to destroy an opponent's nuclear and precision weapon delivery capability. It can also employ nuclear weapons to destroy main attacking groups or significant penetrations and to deny large areas to an opponent.

Note. TC 7-100.2 has been used for all example discussions involving enemy actions.

3-36. With consideration of the threat situation, the CBRN planner advises on the CBRN force employment that best supports the tactical operation. In the offense, the assess function is more important than the functions of protect and mitigate because it provides the commander with early warning of CBRN attacks. Tasks in the assess function allow proactive decision making that enables the tactical commander freedom of action.

Assess

3-37. The CBRN assess function helps commanders decide when and where to concentrate elements of combat power, to include assigned CBRN enablers. CBRN staff elements are integrated with the IPB process and information collection to provide the commander with the information needed to anticipate the enemy's most likely and dangerous courses of action and to provide counteractions that negate CBRN effects.

3-38. In the offense, the CBRN staff identifies threats to the corps and division support and consolidation areas, such as CBRN employment by enemy special-purpose forces and by irregular activities that may interfere with the control of the attack. They assess the vulnerability of critical assets and sites to CBRN, enabling operations in the corps and division deep and close areas. In situations with an active or imminent CBRN environment, they assess risks associated with CBRN conditions and the protective measures required to maintain freedom of action. At the corps and division echelons, the CBRN staff (in coordination with the G-2) assist the G-3 in synchronizing the capabilities of joint, multinational, and national assets in the collection effort. They recommend specific CBRN reconnaissance tasks for corps and division controlled reconnaissance forces and the allocation of CBRN reconnaissance enablers. Information is collected on previous attacks to aid in predictive analysis to select and target the employment of sensors, which further improves the integrated CBRN warning and reporting system. The G-3 synchronizes intelligence operations with combat operations to ensure that all corps and division information collection activities provide timely information in support of operations. The G-3 tasks information collection assets to support the targeting process of decide, detect, deliver, and assess in keeping with the corps and division commander's PIRs.

3-39. During the course of offensive maneuver, CBRN reconnaissance assets assess and characterize sensitive sites, answering PIRs for the commander. R&S platoons that operate forward in close and deep areas locate hazard sites, uncontaminated avenues of approach, and contaminated areas that need to be reported. Planners should also consider the use of special operations CBRN reconnaissance when their unique capabilities are needed.

Note. See ATP 3-05.11 for additional information on special operations CBRN reconnaissance. See ATP 3-11.37 for more information on CBRN reconnaissance and surveillance.

PROTECT

3-40. In the offense, the protect function supports freedom of action against CBRN threats and hazards. Unit movement into attack positions is thoroughly coordinated and planned in detail to preserve surprise. Force concentrations take place quickly and make maximum use of operations security. Units use cover and concealment, signal security, and military deception actions. The attacking force organizes to cope with the environment. This may include attacking across obstacles and rivers, during snow or rain, at night, or on battlefields containing nuclear or chemical hazards. CBRN reconnaissance units support maneuver units throughout their attack and orient on command objectives identified during the IPB process.

3-41. CBRN planners contribute to unit protection by performing vulnerability assessments and ensuring access to critical lines of communications. These assessments provide a list of recommended actions to limit the impact of operating in a CBRN environment. They conduct MOPP analysis and advise on actions that can be taken to protect the force during operations. The warning and reporting system is established, tested, and employed during the offense to collect real-time information on CBRN hazards. CBRN reconnaissance elements detect contamination along routes of advance and monitor lines of communication.

MITIGATE

3-42. During the offense, the CBRN staff supports the lethal targeting process to destroy or neutralize the enemy's CBRN capability and prevent imminent employment against friendly forces. Commanders position available decontamination assets to support their scheme of maneuver. In the close area, thorough decontamination is typically not conducted; however, assets may be required to augment operational decontamination. A consideration for conducting decontamination is to restore combat power. If the operational tempo allows—or if a critical opportunity exists—to consolidate gains through CWMD, the commander may direct maneuver units to allocate combat power to support CBRN site exploitation activities.

3-43. The objective of operational decontamination is to remove enough of the contamination to allow Soldiers to sustain operations. Operational decontamination should begin at the earliest opportunity. Operational decontamination uses two decontamination techniques—MOPP gear exchange and vehicle wash down. The contaminated unit uses its organic equipment to conduct the wash down. Lowering contamination to negligible risk and allowing the reduction of a MOPP level requires a thorough understanding of the risks and rewards associated with decontamination. The resources and time required must be weighed against the advantage gained by the relief from MOPP 4. In addition, CBRN decontamination elements are planned within brigade support areas to enable the rapid reconstitution of combat power and mitigate the spread of hazards within the maneuver space.

HAZARD AWARENESS AND UNDERSTANDING

3-44. During offensive tasks, the CBRN staff supports the commander's decisionmaking process by increasing CBRN awareness and understanding. An integrated reporting system of tactical and technical reports relays actionable information that raises situational awareness. It is critical that reports are prepared and sent in a timely manner to aid follow-on decisions and promote awareness of CBRN hazards. Initial reports (size, activity, location, unit, time, and equipment reports or patrol reports) can be used to report information from the Soldiers on the ground serving as the primary sensors. Units may develop site reports to provide information from deliberate or opportunity targets. CBRN reports are scaled in terms of information and time. CBRN warning and reporting is an information management function that entails collecting and analyzing data from assessments to support operations. This information improves the evaluation and application of access/protect/mitigate. Hazard awareness and understanding uses information gained from assess tasks to identify protection and mitigation tasks.

3-45. The CBRN staff at the division plays a key role in facilitating hazard awareness and understanding as a central point in the warning and reporting system. CBRN 1, 2, and 4 reports move up from reporting units to the CBRN staff. Information is analyzed, and CBRN 3 and CBRN 5 reports are prepared and distributed as required to units in the AO. CBRN staffs use awareness and understanding of CBRN hazard areas to advise the commander on the impacts on the freedom of maneuver.

Note. More detailed information on CBRN warning and hazard prediction can be found in GTA 03-06-008 and TM 3-11.32.

OFFENSIVE PLANNING CONSIDERATIONS

3-46. Operations in CBRN environments require special considerations, regardless of the element of decisive action, the phase of the operation, or the unit involved. Every organization has some basic level of capabilities to assess, protect against, and mitigate CBRN hazards, and every echelon down to the individual level has some basic responsibility to be able to conduct their assigned tasks in CBRN environments.

3-47. The Army echelons its CBRN capabilities to perform different functions. These functions vary with the type of unit, the nature of the conflict, and the number and types of friendly forces committed to the effort. At each echelon, a commander or leader task-organizes available capabilities to accomplish the mission. The purpose of task organization is to maximize different subordinate abilities to generate a combined arms effect. Commanders and staffs work to ensure the distribution of capabilities to the appropriate components of the force to weight the decisive operations and main effort.

3-48. Prior planning allows the commander to exploit the initiative in CBRN environments. If the enemy chooses to employ CBRN to deny terrain and canalize forces, friendly units trained in CBRN defense skills allow the commander to assume risk.

3-49. The tempo of offensive operations requires thorough planning by CBRN staffs to coordinate the challenges of sustaining forces. BCTs plan for site exploitation, the use of contaminated routes for movement, casualty evacuation of contaminated casualties, prolonged operations in CBRN environments, and immediate operational and thorough decontamination of units. The CBRN staff consider the contamination involved, the weather, and METT-TC when providing advice to the commander. For example, a unit that becomes contaminated during offensive operations may continue to fight until the objective is achieved. The commander may give priority for decontamination to artillery units. See appendix D for more information on CBRN staff planning.

3-50. Commanders consider mission variables when designating objectives. Special considerations for CBRN environments that the commander and staff consider within the complementary elements when planning offensive operations include—

- **The scheme of maneuver.** The scheme of maneuver describes how subordinate units relate to each other in time, space, resource, and action. The CBRN staff advises about the best employment options that achieve effective synchronization and complement the maneuver plan. A proper understanding allows CBRN enablers to effectively integrate with maneuver forces. CBRN planners must understand the COCOM's scheme of maneuver to effectively support it. CBRN enablers move under the direction of a supported maneuver force. The CBRN staff can better advise the commander on how CBRN units relate to supported units in time, space, action, and resources by understanding the concept of operations, terrain management, and control measures used. The commander should plan actions for the use of CBRN reconnaissance with the understanding that they are a limited asset. CBRN reconnaissance is placed in time and space to the best advantage of friendly forces based on the enemy templated actions.
- **Deep operations.** Deep operations occur simultaneously with close and consolidation area operations, with echelons above battalion incrementally targeting into deep areas. CBRN staffs must understand the implicit effects of CBRN hazards to the maneuver commander's plan as forces advance into deep areas. When planning fires, the commander must consider intelligence about the enemy's CBRN capabilities and the impacts of hazard areas created from collateral damage. Targeting enemy CBRN storage areas, delivery systems, or industrial areas may have disastrous effects, producing noncombatant casualties and disrupting maneuver. CBRN planners need to advise the commander on the employment of reconnaissance assets in deep areas to provide accurate assessments of hazards and locations of enemy CBRN sites.
- **Reconnaissance and security operations.** Reconnaissance elements may provide early warning of CBRN hazards. All forces make use of the CBRN warning and reporting system to provide early warning of CBRN hazards and allow immediate protection to friendly forces operating in the area. CBRN passive defense measures are taken to protect friendly forces, installations, routes, and actions within a specific area. The CBRN force directly supports these operations with its CBRN R&S assets. Security operations (screen, guard, and cover) provide early warning and protection to the maneuver force. Again, CBRN assets directly support these operations with critical enablers. More information about CBRN reconnaissance, including route, area, and zone, can be found in ATP 3-11.37.
- **Decisive operations and main attacks with shaping operations.** A proper understanding of decisive and shaping operations allows the CBRN force to direct an appropriate priority of effort and priority of support with limited assets. Understanding the threat guides decisions on the priority of support. The priority of effort can be focused around the CBRN functions of assess, protect, and mitigate. The commander must consider the impact on CBRN environments where the enemy is likely to try to fix forces. CBRN planners assist in determining the use of CBRN by the enemy to fix and disrupt maneuver forces. The assigned CBRN brigade resources the corps decisive operations. The assigned CBRN battalion resources the division decisive operation.
- **Reserve operations.** Maneuver commanders at echelon plan to maintain and employ a tactical reserve element to reinforce success or to dynamically react to the threat's initiative. This reserve CBRN force reinforces decisive operations or responds to dynamic threats. CBRN planners need to consider actions to disperse or protect the limited capabilities to preserve their availability as a reserve force. The reserve force maintains the ability to sustain the attack if main effort forces become contaminated as the enemy tries to fix them. Contingencies for the use of the reserve should be part of the corps or division plan. The operation order assigns BE PREPARED tasks to the reserve to aid in its planning and execution. Based on risk, the commander assigns CBRN forces to an appropriate echelon to focus their efforts on assess, protect, or mitigate. These CBRN forces are needed to consolidate gains by further exploiting sensitive sites or conducting decontamination. Operating in protective postures is possible; however, it slow units down.
- **Support area and consolidation area operations.** Planning for support and consolidation area security is a critical role for the CBRN force because these areas have the preponderance of critical assets and infrastructure. Furthermore, the CBRN force enables open ground lines of communication (roads, rail, routes) to provide the movement of sustainment and critical assets to

the close area. Commanders and staffs consider the protection tasks necessary to maintain offensive momentum; for example, CBRN operations. The presence or use of CBRN impacts synchronizing decisive, shaping, and sustaining operations.

3-51. The CBRN planner needs to be able to advise the commander on the expected actions the enemy may take with the employment of CBRN. Possible enemy employments of CBRN include but are not limited to the following:

- Chemical agents employment in AAs, wet gap-crossing sites, disruption zones, and axis of advance.
- Chemical and biological agents employment at airfields, logistics facilities, and mission command centers.
- Tactical nuclear weapons employed against main attacking groups or significant penetrations.

Note. See ATP 3-11.36 for more information on assessing threat capabilities.

CBRN SUPPORT TO OFFENSIVE OPERATIONS

3-52. The four types of offensive operations that apply to the tactical and operational levels of warfare are movement to contact, attack, exploitation, and pursuit. These tasks are designed to defeat and destroy enemy forces and to seize terrain, resources, and population centers. Although the names of these offensive operations convey the overall aim of a selected offensive, each typically contains elements of the other. Corps and division commanders combine these tasks with the forms of maneuver based on their intent and the higher echelon commander's concept of the operation. Forms of maneuver are distinct tactical combinations of fire and movement with a unique set of doctrinal characteristics that differ primarily in the relationship between the maneuvering force and the enemy. The six forms of maneuver include envelopment, turning movement, frontal attack, penetration, infiltration, and flank attack. Offensive operations are characterized by surprise, concentration, tempo, and audacity.

Note. See FM 3-90-1 for additional information on offensive operations, their forms and characteristics, and forms of maneuver.

3-53. The organization of friendly forces for each offensive task is slightly different and is based on the concept of operation to defeat the enemy. The offensive tasks of movement to contact and the attack typically involve a forward security force or advanced guard, a main body formation that could have flank and rear security, and a reserve or follow-on force. Friendly forces organize a self-sufficient main body for pursuit and exploitation tasks. Mission command elements are positioned forward during exploitations and pursuits, which increases their risk to enemy contact. Corps and division commanders reinforce their decisive operations with additional combat power, such as reconnaissance, fires, mobility, attack aviation, air defense, and sustainment. Divisions and BCTs must consider the enemy's ability to counterattack during the offense and may employ additional security forces on exposed flanks.

3-54. CBRN support to the offense includes the simultaneous application of the assess, protect, and mitigate functions guided by hazard awareness and understanding. CBRN commanders at echelon with the supported unit staffs recommend the allocation and command relationship of CBRN assets and their capabilities to enable movement and maneuver of main body elements according to the concept of the operation, scheme of maneuver, and framework of the prevailing offensive task.

3-55. Brigade and battalion CBRN headquarters advise supporting maneuver commanders on the further allocation and tactical assignment of subordinate CBRN companies that have task-organized reconnaissance, decontamination, site exploitation, and bio-detection assets. These headquarters enable the operations process and the essential sustainment functions during the offense, to include planning, preparation, liaison, and integration activities.

3-56. CBRN R&S assets support the offense as part of a larger maneuver or security force or as a specialty platoon directed by a maneuver headquarters. These assets—which are normally organized, allocated, and assigned at the platoon level—primarily perform information collection tasks directed at CBRN PIRs and their relating NAIs. They are also capable of performing general reconnaissance, security, and tactical enabling tasks to support the offense, such as route reconnaissance and the forward passage of lines. Otherwise, they augment security operations to provide early warning of CBRN threats and hazards to protect the decisive operation or main body in the close or support areas. This includes monitoring key lines of communication. CBRN reconnaissance assesses suspected CBRN storage areas that have or can be used against friendly forces or populations. During pursuit operations, the enemy may attempt to utilize CBRN to break contact; therefore, it is best to increase CBRN reconnaissance capabilities to maintain pressure on retrograde forces.

Note. See ATP 3-90.40 for more information on the employment considerations of CBRN technical enablers during combined arms CWMD.

3-57. Decontamination assets support the offense by providing mobility along contaminated portions of key terrain or by reconstituting essential combat power that the enemy denied or neutralized through a CBRN attack. They can augment the organic operational decontamination capability of maneuver forces or perform thorough decontamination of equipment at specific locations. These assets are organized at the platoon or squad level and are allocated under a CBRN company. Commanders can further organize decontamination assets to support battalion or battalion level equipment decontamination; however, these tasks take several hours of preparation and execution, which typically occur during the consolidate gains portion of the offense. Thorough decontamination is a reconstitution effort that is conducted after the battle.

3-58. CBRNE assets support the offense by conducting time-sensitive information collection and site exploitation on suspected WMD or TIM sites across the AO, but these typically occur in a consolidation area established by a division headquarters.

Support to Offensive Operations (Attack)

3-59. Attacks incorporate coordinated movement supported by fires and may be hasty or deliberate, depending on the time available for the operations process. An attack differs from a movement to contact; during an attack, commanders know at least part of an enemy's disposition to better orchestrate the warfighting functions and to more effectively concentrate the effects of combat power. Commanders normally organize the attacking force into a security element, a main body, and a reserve, which are all supported by some type of sustainment organization.

3-60. Commanders organize the main body to conduct the decisive operation and necessary shaping operations. The decisive operation is focused toward a decisive point that can consist of the defeat of enemy forces or the seizure of decisive terrain. Shaping operations create windows of opportunity for executing the decisive operations and receive the minimum combat power necessary to accomplish their missions. Commanders use their reserves to exploit success, defeat enemy counterattacks, or restore momentum to a stalled attack. The commander resources sustaining operations to support the attacking force while further organizing its internal sustainment capability. For example, maneuver battalions organize their internal sustainment assets into combat and field trains while the brigade support battalion controls sustaining operations to displace as forward as required to shorten supply lines.

3-61. Commanders use necessary control measures to coordinate and synchronize the attack. Units conducting the attack are assigned an AO within which to operate and further designate subordinate AOs for units of battalion size or larger. Phase lines (PL) allow the commander to coordinate and synchronize movement and supporting fires toward the enemy. PLs are typically associated with clearly recognizable terrain features, such as roads, streams, wadis, or inter-visibility lines. A line of departure or line of contact designates areas beyond which friendly and enemy forces will make likely or imminent contact, and existing PLs typically receive these secondary designations. An objective is used to designate key or decisive terrain with known enemy concentrations of significant value toward accomplishing the mission. If necessary, a commander can use an axis of advance or a direction of attack to further control maneuver forces. Short of the line of departure, a commander may designate AAs and attack positions where units prepare for the offense. Beyond the line of departure, the commander may designate direct fire control measures, fire support coordination measures, final coordination lines, assault positions, attack by fire and support by fire positions, or a probable line of deployment.

3-62. Figure 3-5 and figure 3-6, page 3-14, provide tactical COA graphics to illustrate the integration of CBRN forces with maneuver during an offensive attack. In figure 3-5, the hazard assessment platoon from the hazard response company is task-organized to Task Force 1 (TF-1), which has objective four (OBJ 4), a suspected radiological site, in its area of operations. The CBRN R&S platoon with the armored reconnaissance squadron is an organic asset. The graphic is similar to the course-of-action sketch example in FM 6-0, which includes task-organization icons found in ADP 1-02.

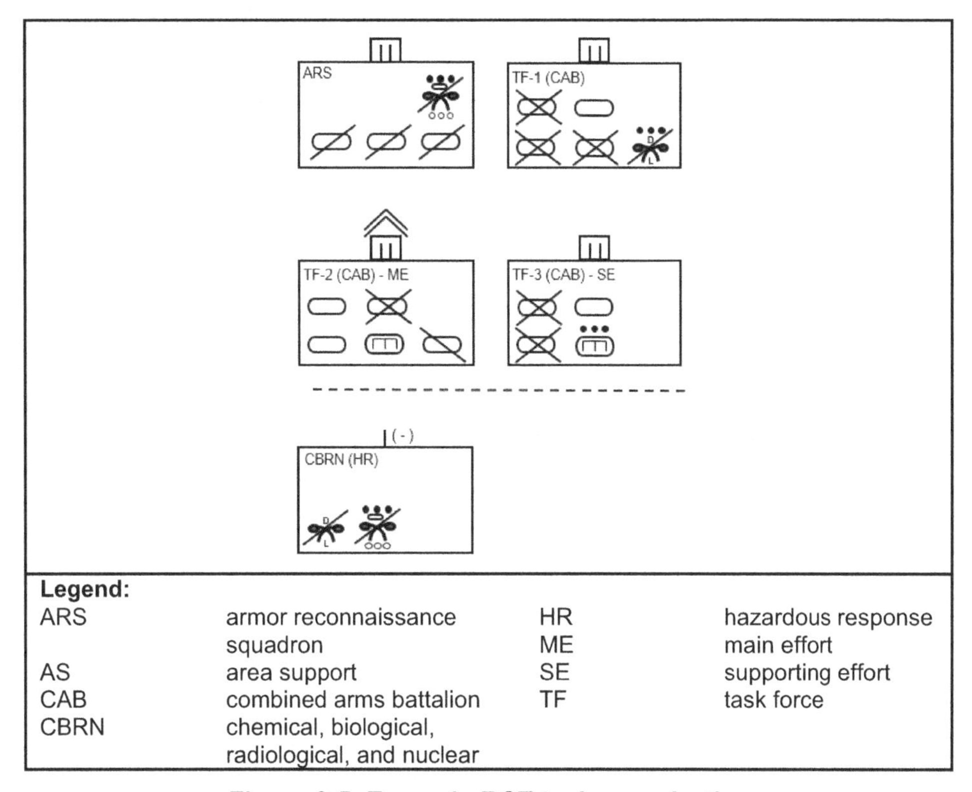

Figure 3-5. Example BCT task organization

3-63. In figure 3-6, an armored brigade combat team (ABCT) conducts a frontal attack along three directions of attack to defeat a task-organized enemy battalion in an area defense on the nearside of PL BLUE to enable the division's follow-on operations past the brigade's limit of advance. The enemy consists of a motorized infantry battalion task-organized with two motorized infantry companies, a mechanized infantry company, a tank platoon, and a towed (medium) artillery battery. The OE includes CBRN threats and hazards with an enemy most likely COA that involves the use of chemical attack to reinforce its defense along canalized terrain on the right side of the sketch. The OE also has an urban area (DODGE CITY) with a radiological facility that the ABCT must isolate and secure. The technical enablers assigned to the task force assess the facility. Figure 3-5 is a depiction of the task organization used in figure 3-6. It uses task-organization composition symbols to increase understanding of how an ABCT and all of its organic enablers organize for the attack with the attachment of a CBRN hazard response company. The organization of the brigade engineer battalion (BEB) and battlefield surveillance brigade are not depicted and include all organic assets except for one combat engineer company and an additional combat engineer platoon allocated to the combined arms battalions. The hazard response company is attached to the BEB; it retains the heavy R&S platoon and detaches one of the hazard assessment platoons to TF-1. The BCT's organic CBRN R&S platoon is integrated into the armored reconnaissance squadron and is conducting a screen, with sensors focused on the CBRN named area of interest at canalizing terrain tied into enemy obstacles along direction of attack NIKE.

Note. See FM 6-0 for more information.

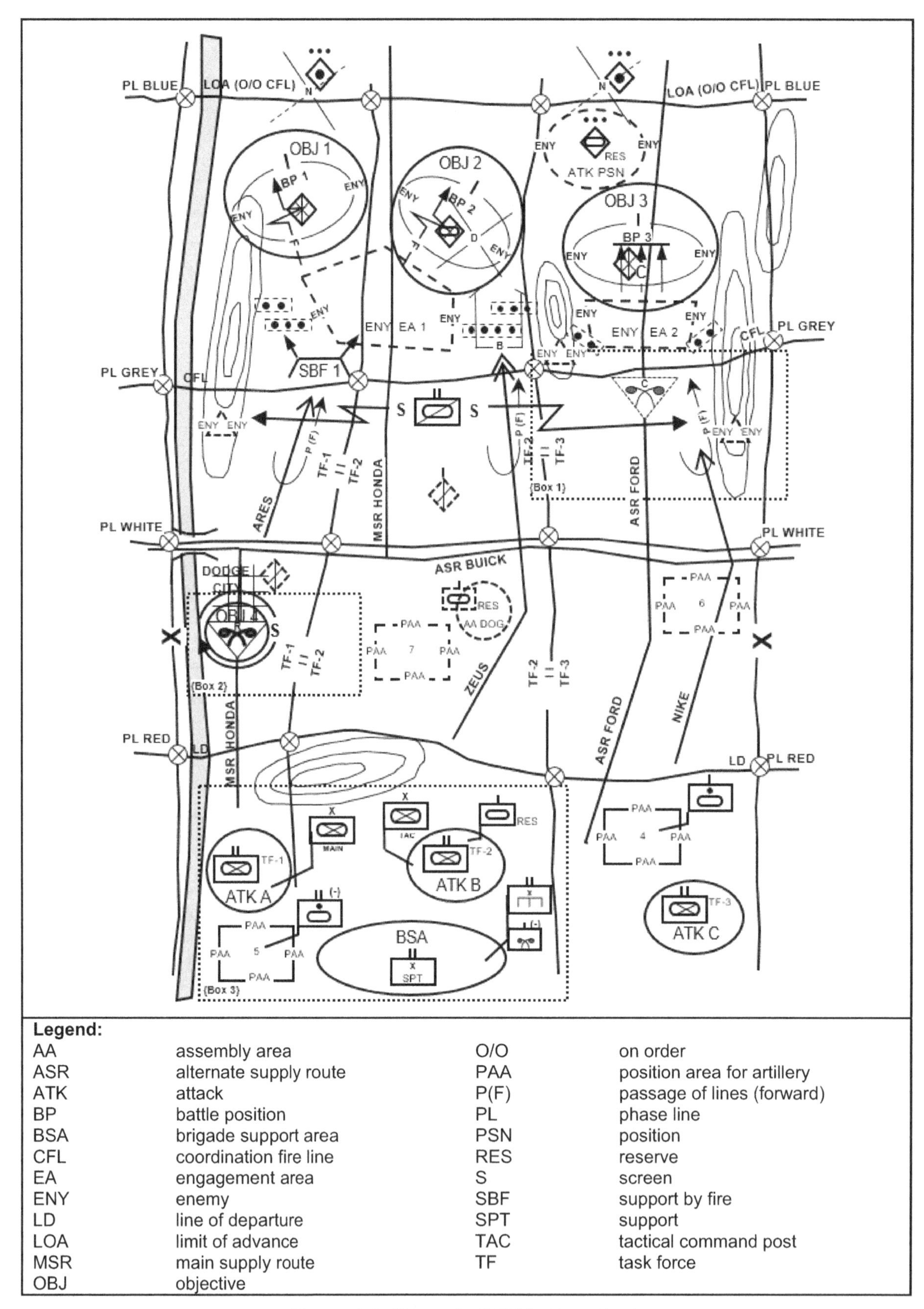

Legend:

AA	assembly area	O/O	on order
ASR	alternate supply route	PAA	position area for artillery
ATK	attack	P(F)	passage of lines (forward)
BP	battle position	PL	phase line
BSA	brigade support area	PSN	position
CFL	coordination fire line	RES	reserve
EA	engagement area	S	screen
ENY	enemy	SBF	support by fire
LD	line of departure	SPT	support
LOA	limit of advance	TAC	tactical command post
MSR	main supply route	TF	task force
OBJ	objective		

Figure 3-6. Example BCT in the attack

3-64. The COA depicted in figure 3-6 involves an armored reconnaissance squadron and two combined arms battalions (TF-1 and TF-3) conducting a shaping operation ahead of the decisive operation conducted by the third combined arms battalion (TF-2). The armored reconnaissance squadron is task-organized with its three organic cavalry troops and one heavy CBRN R&S platoon and conducts a screen along PL GREY to deny enemy reconnaissance and provide freedom of maneuver for follow-on operations. On order, the armored reconnaissance squadron conducts a forward passage of lines along PL GREY to move all three task forces forward while maintaining contact with the enemy. TF-1 secures objective (OBJ) four (OBJ 4), a radiological facility in DODGE CITY, and simultaneously moves along direction of attack ARES, conducts forward passage of lines with the ARS, and then establishes support-by-fire position (SBF)-1 to fix enemy (ENY) on objectives one and two (OBJ 1 and OBJ 2) to enable the decisive operation. The supporting effort, TF-3, moves along direction of attack NIKE, conducts forward passage of lines with the ARS, and then clears a motorized infantry company on OBJ 3 to prevent the enemy from reinforcing OBJ 2. The decisive operation and main effort, TF-2, moves along direction of attack ZEUS, conducts forward passage of lines with the ARS, and then conducts a task force breech to destroy one mechanized infantry company on OBJ 2 to enable division follow-on operations past of the limit of advance. The BEB conducts area security operations up to PL RED to enable the decisive operation. The brigade reserve, a tank company, moves along direction of attack ZEUS and establishes AA DOG. The priority of commitment for the reserve is to reinforce the decisive operation. The fires battalion initially provides priority of fires to the ARS, TF-3, and then TF-1 from position areas for artillery four and five (position area for artillery [PAA] 4 and PAA 5). On order, the fires battalion displaces to PAA 6 and PAA 7 to provide priority of fires to TF-2 (main effort) followed by TF-3 and then TF-1. High-payoff targets are enemy armor, mechanized infantry forces, and indirect fire systems. The coordination fire line is initially PL GREY and on order shifts to PL BLUE aligned to the forward passage of lines operation. The BSB conducts sustaining operations from the brigade support area using main supply route (MSR) HONDA and alternate supply routes FORD and BUICK to initially sustain the shaping operation. On order, the BSB sustains the decisive operation along the MSR and alternate supply routes followed by the shaping operation with priority to the main effort and then the supporting effort.

3-65. The next series of figures provide expansions of specific areas of figure 3-6 to further explain the integration of maneuver and CBRN forces. Figure 3-7, page 3-16, shows an expansion of Box 1 from figure 3-6 to explain how the heavy CBRN R&S platoon integrates with the ARS. The armored reconnaissance squadron initially conducts route, zone, and counter reconnaissance along the three directions of attack and respective task force boundaries past PL WHITE. On order, the armored reconnaissance squadron shifts to its decisive operations of conducting forward passage of lines with all three task forces while maintaining contact with the enemy. Figure 3-7 depicts the movement of a cavalry troop that conducts route reconnaissance along direction of attack NIKE and zone reconnaissance past PL WHITE in the battalion boundary for TF-3. This shows a heavy CBRN R&S section from its parent platoon integrating with other armored reconnaissance sections from two organic platoons under a cavalry troop. A cavalry troop typically infiltrates by sections, controlled by platoons, with observation posts as final march objectives. The reconnaissance and surveillance of NAIs should occur through redundant observation posts to prevent the loss of surveillance from the compromise of a single position. Organic cavalry sections normally lead the route, zone, and counter reconnaissance to set conditions for the movement and positioning of enabling reconnaissance elements. Initially, cavalry sections infiltrate along independent routes to establish forward observation posts that support tactical enabling tasks and surveillance of assigned NAIs to meet the commander's PIRs. The heavy CBRN R&S section subsequently moves along a separate infiltration route and sets at observation post ten (OP 10) to conduct surveillance on named area of interest 5 in support of the ABCT commander's CBRN PIRs.

3-66. This graphic also enables a discussion of fires in support of maneuver to suppress or destroy enemy observation posts (counter-reconnaissance). The cavalry troop requests fires on target reference points four and five (TRP 4 and TRP 5), which are suspected enemy positions to enable their movement toward friendly observation posts and to deny enemy surveillance of the friendly scheme of maneuver. If these sections identify known enemy observation posts, they adjust fire from the target reference points to engage the enemy without the use of direct fire, which supports continued infiltration and concealment during the reconnaissance fight.

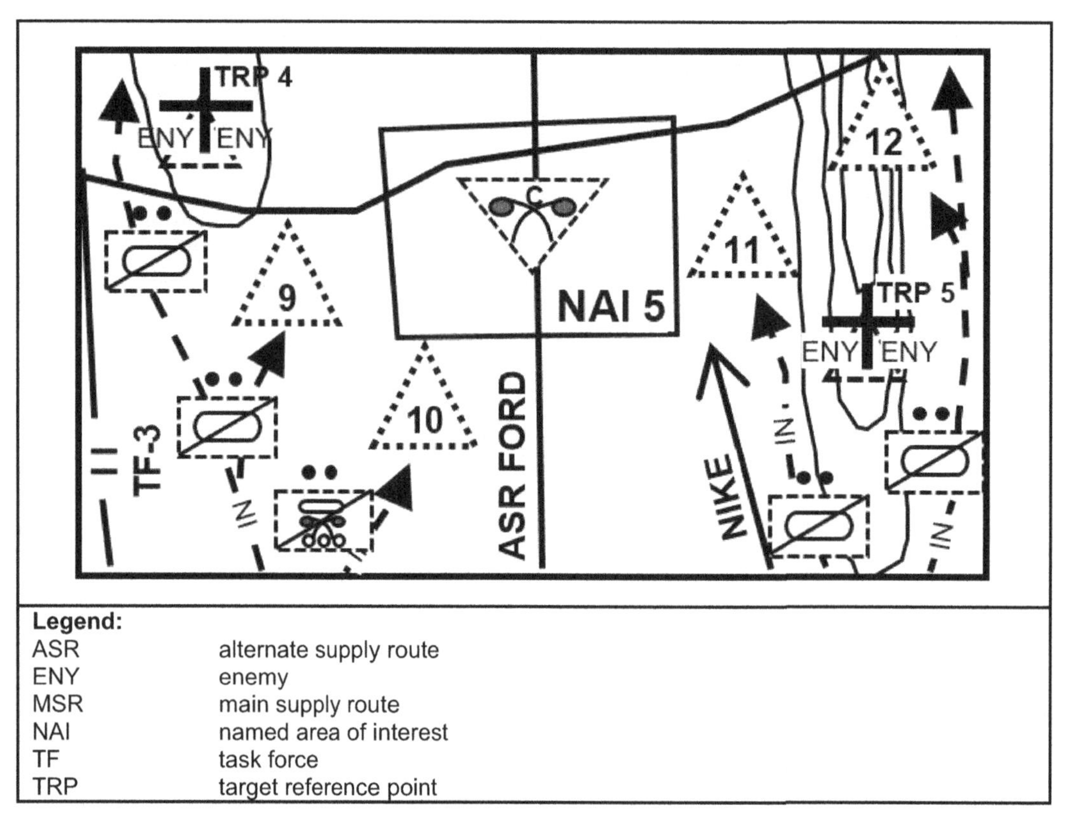

Figure 3-7. Example COA for a chemical NAI

3-67. Figure 3-8 shows an expansion of Box 2 from figure 3-6, in which a hazard assessment platoon integrates with a combined arms battalion to secure, assess, and characterize objective four (OBJ 4), a radiological facility in DODGE CITY. The combined arms battalion, Task Force 1 (TF-1), utilizes minimum combat power to secure the facility while it simultaneously commits its remaining combat power to establish SBF-1 to enable the brigade's decisive operation well past OBJ 4. The hazard assessment platoon does not conduct its mission without the integration of a mechanized infantry security element. The situation depicted in figure 3-8 shows the threat of a suspected enemy reconnaissance element of unknown size operating in the vicinity of DODGE CITY. This is a threat to OBJ 4 and to the movement of TF-1 past DODGE CITY. TF-1 organizes for its operation by assigning OBJ 4 to a mechanized infantry company task-organized with the hazard assessment platoon. The remainder of the TF-1, composed of a mechanized infantry company team, a tank company team, and a pure mechanized infantry company, bypasses DODGE CITY along axis of advance SWORD.

3-68. In figure 3-8, a mechanized infantry company secures the radiological facility within its company boundary with two platoon size battle positions on the far side of OBJ 4. TF-1 will likely lead the movement from its attack position with the company assigned to OBJ 4 as a diversion to draw enemy reconnaissance toward the radiological facility. However, it can conduct both movements simultaneously as long as the battalion coordinates the simultaneous maneuver of multiple companies through appropriate control measures. After the company assigned to the city establishes its two battle positions, it moves the hazard assessment platoon with a platoon size security element and establishes AA CAT on the near side of OBJ 4. The mechanized infantry platoon then establishes an inner cordon around the facility to allow the hazard assessment platoon to further assess and characterize the suspected radiological material.

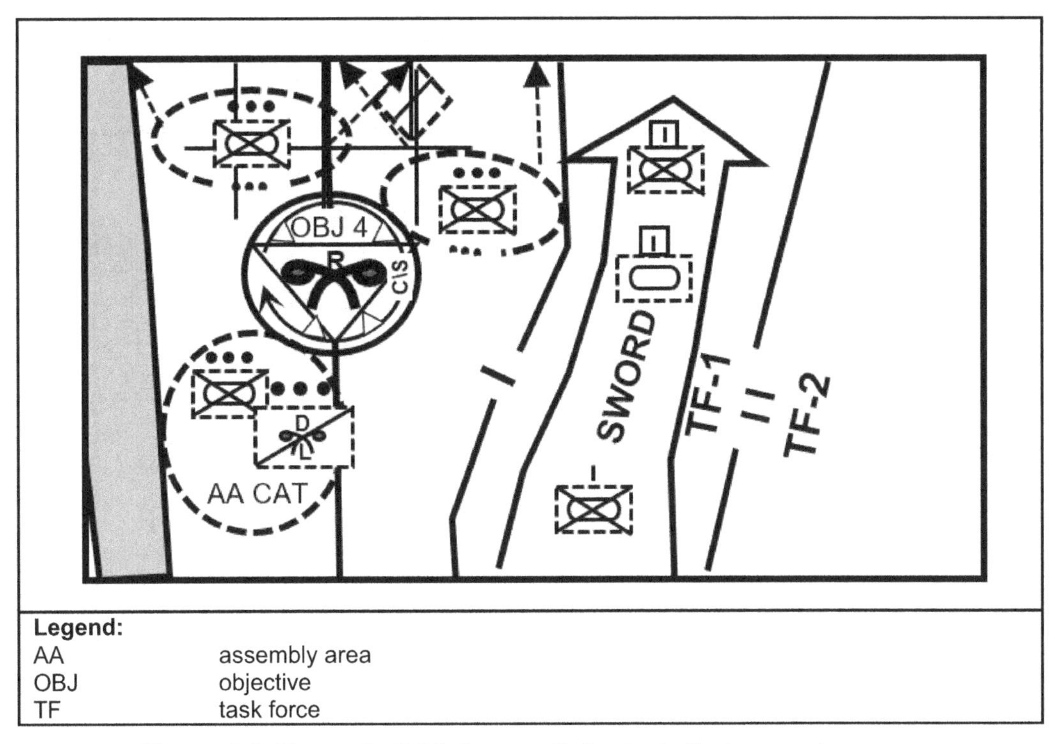

Figure 3-8. Example COA for a radiological site in an urban area

3-69. Figure 3-9, page 3-18, shows an expansion of Box 3 from figure 3-6, in which a hazard response company integrates with a BEB to conduct area security up to PL RED in the ABCT's support area. During the offense, support areas are usually organized with clusters of AAs and attack positions connected through interior lines of communication, such as unimproved roads bounded by entry control or check points. The BEB enables maneuver and sustainment operations by providing command and control of the support area while adhering to the five fundamentals of area security operations. This includes the ability to provide early and accurate warning, provide reaction time and maneuver space, orient on the force to be protected, perform continuous reconnaissance, and maintain contact with the threat.

3-70. The CBRN threats and hazards depicted in figure 3-8 show the existence of radiological material in DODGE CITY and the potential use of chemical munitions by the enemy. In figure 3-9, the BEB uses its assigned combat power to establish entry control points one and two (EC 1 and EC 2) to prevent mounted access to interior roads by threats and local civilians. The BEB is responsible for protecting the brigade main command post and brigade support area, which will continue to operate in the support area throughout the attack. The hazard response company heavy CBRN R&S platoon conducts route reconnaissance along the MSR and interior lines to provide early warning of route contamination by friendly forces or threat actions. The BEB has identified a decontamination point near EC 1 and has assigned the hazard response company's hazard assessment platoon with the on order task of establishing and operating an operational decontamination point to mitigate contamination from potential indirect fires in the support area. The BEB is responsible for securing and augmenting the decontamination point with its own organic decontamination assets and security because the decontamination platoon does not have sufficient security or personnel assets for sustained operations. In the situation depicted in figure 3-9, the decontamination point is located just outside the area bounded by the entry control points, which can provide mutual security for the decontamination platoon.

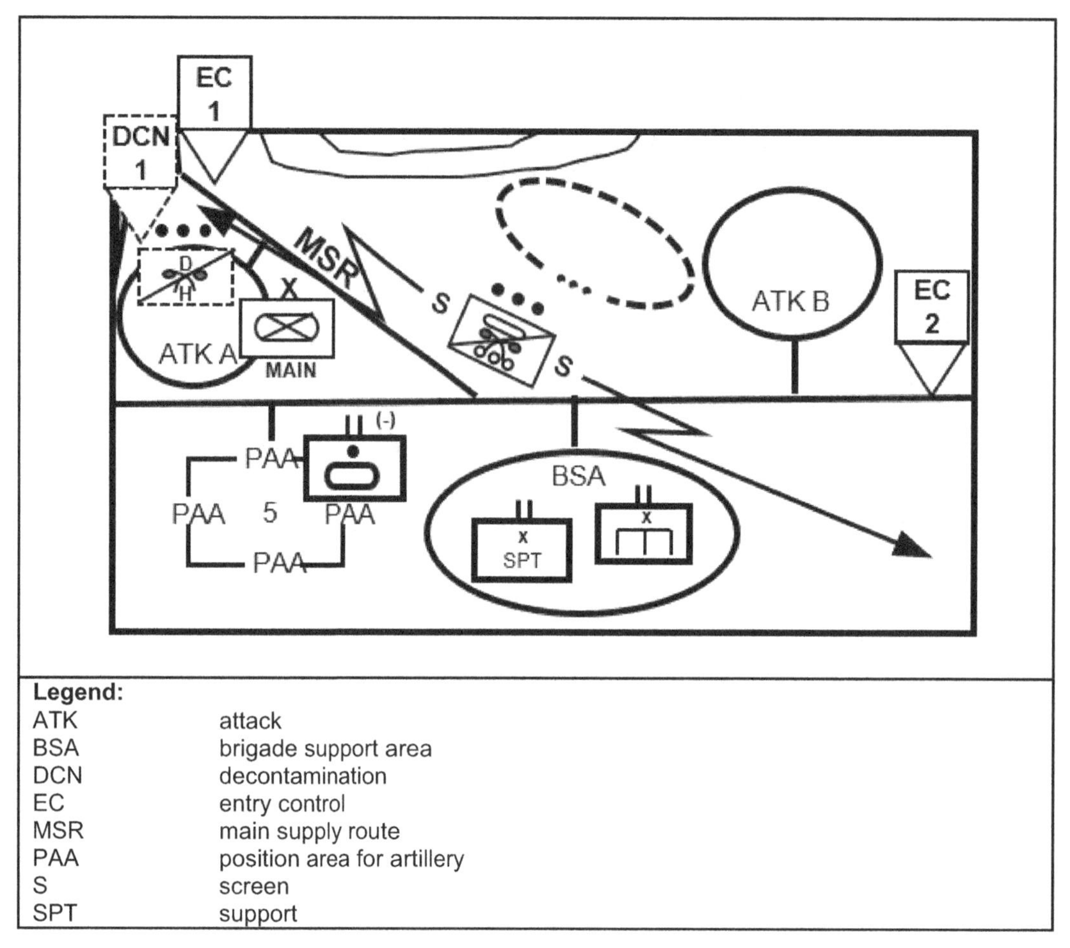

Figure 3-9. Example COA for a support area

CONSOLIDATE GAINS

3-71. The consolidation of gains is an integral part of the offense and is essential to retain the initiative over determined enemies and adversaries. The emphasis is on early and effective consolidation activities to enable success and achieve lasting favorable outcomes in the shortest time span. This encompasses minimum essential stability tasks to allow the maneuver commander options while isolating CBRN hazards and securing WMD to prevent them from falling back into the hands of the enemy. CBRN units must be prepared to quickly assess abandoned enemy CBRN caches in the open and in underground facilities after the offense to consolidate gains. Planners must consider security requirements for WMDs to prevent enemy acquisition and usage. CBRNE companies, biological surveillance companies, and area support companies consolidate gains within corps and division consolidation areas as the forward line of troops advances in the offense. Area support companies decontaminate aerial ports of debarkation and seaports of debarkation facilities to retain operational reach.

SECTION II—CBRN CAPABILITIES IN DEFENSIVE OPERATIONS

OVERVIEW

3-72. CBRN personnel contribute to the defense through protection tasks and support to large-scale decontamination efforts. The CBRN defense plan is the result of CBRN staff conducting threat assessment to understand enemy CBRN capabilities. Next, the staff plan protection requirements by performing vulnerability assessments. The vulnerability reduction measures mitigate weaknesses in the defense against the CBRN threat. These assessments provide a list of recommended actions, including protection measures for personnel, supplies, facilities, and equipment from contamination, collective protection (COLPRO), and contamination mitigation based on the threat situation.

3-73. Defensive operations are conducted to defeat an enemy attack, gain time, economize forces, and develop conditions favorable for offensive or stability tasks. There are three types of defensive operations—area defense, mobile defense, and retrograde. Although the names of these defensive tasks convey the overall aim of a selected defense, each typically contains static and mobile elements.

3-74. Defending forces must use CBRN defensive principles to protect the force, avoid contamination where possible, and conduct decontamination to preserve combat power. The following are some characteristics of the defense that are of concern for CBRN operations:

- **Flexibility.** Flexible plans anticipate enemy actions and allocate resources accordingly. Commanders must be prepared to defend against the enemy use of CBRN and plan protection measures and necessary resources to disrupt the enemy desired effects.
- **Preparation.** Defenders use available time before an attack to prepare. CBRN staffs understand considerations of the environment to improve the survivability of the unit. Examples of improving survivability of a unit include: improving the condition and capability of existing buildings or fortifications against CBRN hazards, using COLPRO, and covering or dispersing supplies as preparation measures.
- **Security.** Security tasks provide early warning and preserve combat power. The integration of CBRN detection systems with the whole spectrum of early warning systems provides increased situational awareness and security.

Threat Overview

3-75. The threat overview describes how threat forces organize again U.S. forces in the defense. This allows CBRN forces to understand enemy potential courses of action.

3-76. The threat uses the offense to carry the fight to its adversary to impose its will and achieve significant tactical or operational objectives. Therefore, enemy commanders seek to create and exploit opportunities to take offensive action when possible. The threat decentralizes control through specific mission tasks for subordinate tactical commanders that are tied to nested purposes and which further support the overall goals of the offensive operation. In this way, subordinate commands may continue to execute their missions without direct control by a higher headquarters. The threat also employs battle drills at higher tactical levels to further simplify coordinated action without detailed planning. This increases the enemy's reaction time and ability to exploit the initiative. Enemy commanders use minor, simple, and clear modifications to thoroughly understood and practiced battle drills to adapt to ever-shifting conditions.

3-77. During the offense, enemy forces typically attempt to mask the location of their main effort with multiple fixing attacks on the ground while using fires to disrupt critical nodes and to isolate the forward combat elements of their adversary. Critical nodes that the threat attempts to target include but are not limited to command posts, radars, and fire direction centers. Enemy forces seek to reinforce success, massing capabilities at a vulnerable point to achieve large force ratios to enable a rapid penetration of the adversary's defenses. The enemy uses mobile forces to rapidly exploit penetrations to maximize the potential depth of their offense, making their adversary's defense untenable. Threat forces can have advantages in volume and range of fires so they can simultaneously mass fires on the point of penetration. These advantages enable the threat to fix other elements on the adversary's forward line of troops and to target mission command and logistic nodes along the depth of the defense. Threat forces employ intelligence, surveillance, reconnaissance, electronic and information warfare, special-purpose forces, and chemical weapons to enable its offensive operations.

3-78. Peer threats are able to conduct large-scale offensive action with tactical groups, divisions, and brigades. Typical offensive actions include an attack or a limited-objective attack. The general forms of attack are integrated or dispersed. An integrated attack is employed when the threat has overmatch with its adversary; a dispersed attack is used when the threat is disadvantaged. Integrated attacks are generally characterized by simultaneously focusing on critical nodes while fixing the majority of its adversary's forces in place with the minimum necessary combat power. After the threat has fixed the majority of its adversary, it employs flank attacks and envelopments to further exploit the offense. Dispersed attacks are characterized by an increased emphasis on critical nodes while conducting simultaneous attacks at multiple, dispersed locations. The threat operates with smaller, independent subordinate elements that rapidly move from separate locations and mass at the last possible moment. Limited-objective attacks include spoiling attacks and counter attacks to preempt the initiative of an adversary.

3-79. The threat organizes for the offense with enabling and action forces. With all forms of the offense, enabling forces include fixing, assault, and support forces. The most common type of action force is the exploitation force. Fixing forces prevent an adversary's defending forces from interfering with the actions of the assault or exploitation force. Maneuver forces, precision fires, air defense and antiarmor ambushes, situational obstacles, chemical weapons, and electronic warfare can be used as a fixing force. The typical fixing force is a maneuver force enabled by support forces that add combat power to the mission task of fixing an adversary in place.

3-80. At the brigade tactical group level, the threat commander employs one or more assault forces to destroy a particular part of an adversary's defense or to seize key positions. Support forces are combat or combat-support forces that employ attack-by-fire, indirect fires, close air support, or mobility for other enabling or action forces. As the primary action force, the exploitation force must be capable, through inherent capabilities or positioning relative to an adversary's defense, of destroying the target of the attack. It must be capable of penetrating or avoiding significant defensive positions and attacking an adversary's support infrastructure before it has time to react. An exploitation force ideally possesses a combination of mobility, protection, and firepower that permits it to reach the target with sufficient combat power to accomplish the mission. These forces can typically range from a task-organized tank brigade to widely dispersed groups of special-purpose forces. Based on the form of offense, multiple exploitation forces can move dispersed and mass at points of penetration or form a singular combat force separated by space and time behind assault forces.

3-81. If the strategic or operational situation warrants the use of CBRN capabilities, a peer threat typically uses a more substantial volume to support its offensive operations in comparison with its defensive operations. Due to the volume requirements to deliver WMDs to achieve desired effects, CBRN planners should assess positions of enemy multiple rocket launchers, 152-millimeter artillery batteries, and theater ballistic missiles within range of friendly close and support areas. CBRN support to the offense is more likely because an adversary would normally be in a defensive posture where the locations of critical nodes or significant positions are more defined or known due to the static nature of the defense. As previously stated, the threat's use of nonpersistent or persistent chemical agents, rather than biological or nuclear, is more likely in the offense and defense because enemy forces can anticipate and control the effects.

3-82. It is common for enemies to mix chemical rounds with high-explosive (HE) rounds to achieve desired effects. In the offense, the enemy may use chemical agents to restrict the use of terrain, especially key points along an adversary's lines of communication. Nonpersistent agents are suitable for use against targets on an axis of advance the enemy intends to exploit. Their most likely role is to prepare the way for an assault force or exploitation force, especially when the adversary's positions are known in detail. They can also be used against civilian population centers to create panic and a flood of refugees. Persistent agents are suitable against targets the enemy cannot destroy by conventional or precision weapons, such as targets that are too large or targets that are located with insufficient accuracy for attack by other than an area weapon. Persistent agents can neutralize such targets without a pinpoint attack. The enemy may target bypassed pockets of resistance with persistent agents, especially those that pose a threat to an exploitation force. If the threat can determine possible AAs for its adversary's counterattack forces, it will likely use persistent agents to target those locations. Finally, the threat could use persistent agents deep within its adversary's rear and along troop flanks to protect advancing units. Typical target options for all types of chemical agents when the threat is conducting offensive operations are troop concentration areas, headquarters, and artillery positions. The threat could use chemical attacks against such targets simultaneously throughout an adversary's defense. These chemical attacks combine with other forms of conventional attack to neutralize an adversary's nuclear capability, command and control systems, aviation, and logistics facilities. Delivery options include tube artillery, multiple rocket launchers, surface-to-surface missiles, and aircraft.

3-83. A peer threat has the same considerations in the offense as it does in the defense for the release authority and the employment of biological and radiological capabilities. If the release authority for nuclear weapons has approved the use of tactical weapons, a peer threat may use a nuclear attack in coordination with nonnuclear fires to destroy an adversary's main combat formations, command and control systems, and precision weapons. They can be used to target and destroy an adversary's defense to set conditions for an exploitation force. In those situations, the threat may plan high-speed air and ground actions to exploit the nuclear attack. As an adversary withdraws, the threat may use a subsequent nuclear attack on choke points, where retreating forces present lucrative targets.

3-84. The CBRN planner needs to advise the commander on expected actions the enemy may take with the employment of CBRN against U.S forces in the defense. Examples of enemy CBRN use include—

- The use of CBRN to prevent the strike force from being committed in a mobile defense.
- Persistent chemical strikes to isolate and divide portions of maneuver space.
- The use of CBRN to create simultaneous dilemmas, especially in support areas.
- Nonpersistent chemical strikes to degrade friendly forces at the point of penetration.

Note. See TC 7-100.2 for more information.

PROTECT

3-85. Unit survivability is critical to the success of defensive operations, regardless of the form of the operation. Once again, the protection function provides the commander with freedom of action by preserving subordinate unit capabilities so that the maximum combat power can be applied at the desired time and location. Criticality, vulnerability, and recuperability are some of the most significant considerations for the commander in determining priorities for the protection function. Because defending units are often in fixed and concentrated positions, they increase their vulnerability to CBRN threats and hazards. All units have an inherent responsibility to improve the survivability of their own fighting positions, AAs, and bases. The defending force BCTs occupy their respective AOs as soon as possible to maximize preparation time for defensive positions and obstacles. Survivability operations enhance the ability to avoid or withstand hostile actions by altering the physical environment. Units accomplish this through four tasks: constructing fighting positions, constructing protective positions, hardening facilities, and employing camouflage and concealment. Effective tactical dispersion is an important aspect that commanders must consider; however, massing effects is a principal characteristic of the defense that causes a necessary balance against excessive dispersion. Commanders should consider more frequent displacement of units in close and support areas to prevent identification and acquisition for enemy CBRN strikes in the defense.

3-86. CBRN staffs play a greater role in the defense through the protection function by performing a larger volume of vulnerability assessments. They advise the commander on the employment of CBRN passive defense measures (detection and warning equipment, IPE, COLPRO) to protect personnel, equipment, and facilities from CBRN/TIM effects. The commander positions forces and installations to avoid congestion but does not disperse them to the extent that there is a risk of defeat in detail by an enemy that is just employing conventional munitions. Protecting critical assets, sites, and lines of communication in the corps consolidation area and in division support areas require a sufficient complement of CBRN units to provide versatility in procedures and to increase overall survivability. CBRN staffs must consider how to best maintain freedom of maneuver for reserve or strike forces and apply sufficient priority in the CBRN defense plan. These forces are highly valued, lucrative targets for the enemy to attack with their CBRN capabilities. Considerations include establishing for protective posture status, tactical dispersion, and alternate routes to prevent cross contamination.

ASSESS

3-87. The CBRN assess function directs information collection to enable targeting of the enemy's CBRN capability and to further identify when, where, and with what strength the enemy will attack. These activities further support CBRN active defense measures for friendly forces. Commanders continuously refine the intelligence picture of enemy forces throughout their area of interest as part of deep operations. Corps and divisions focus on the timely, accurate identification of high-payoff targets during defensive preparations. Constant surveillance of the AO and effective reconnaissance are necessary to acquire targets and to verify or evaluate potential enemy courses of action with respective CBRN capabilities. In the close area, commanders use intelligence products to identify probable enemy objectives and approaches. From those probable objectives and approaches, NAIs and targeted areas of interest can be developed. The commander must also examine the enemy's capability to conduct air attacks, insert forces behind friendly units, and employ CBRN capabilities to shape the battle to their advantage.

3-88. The CBRN staff and their reconnaissance assets assist corps and division level intelligence (G-2) sections during the preparation phase to complete information collection integration and synchronization. Corps and divisions rely on joint and national systems to detect and track targets beyond their organic capabilities. Organic and assigned information collection assets identify friendly vulnerabilities and key defensible terrain. A division headquarters conducts periodic information collection of unassigned areas to prevent the enemy from exploiting these areas to achieve surprise. They also use available reconnaissance, surveillance, and engineer assets to study the terrain. By studying the terrain, the commander tries to determine the principal enemy and friendly heavy, light, and air avenues of approach. The commander determines the most advantageous area for the enemy's main attack and conducts other terrain analysis of observation and fields of fire, avenues of approach, key terrain, obstacles, and cover and concealment (OAKOC). During planning, BCT commanders establish PIR and assign available CBRN collection assets to fulfill these requirements. CBRN reconnaissance elements are positioned by the defending unit to gain critical information, which directly supports hazard awareness and understanding. The commander should employ CBRN reconnaissance and surveillance elements along movement routes and at potential choke points. CBRN reconnaissance units can also provide the security force commander with versatility and an increased capacity for surveillance. During the execution phase of the defense, CBRN reconnaissance should focus on determining the status (clear or contaminated) of passage lanes, which redundantly support the withdrawal of a security force or forward movement of a strike force. CBRN reconnaissance units locate and mark contaminated routes and help identify new clear routes.

3-89. The defense allows for a more robust and reliable CBRNWRS to quickly anticipate and react to the enemy's use of CBRN capabilities. Because the threat will likely mix CBRN and conventional munitions during precision or area strikes, an effective reporting system allows commanders to fully understand an enemy commander's COA. In the preparation and execution phases of the operation, the CBRN staff recommends adjustment and works through the echeloned operations officer (G-3) to task subordinate units for additional active and passive sensors to enable timely and accurate reporting.

MITIGATE

3-90. The focus of the mitigate function in defensive operations is to quickly restore combat power if the threat employs CBRN capabilities. The CBRN staff assists the targeting section at the corps and division echelon to adjust the attack guidance matrix to exploit opportunities and target those CBRN capabilities that the enemy commander requires most. In general, corps deep operations occur beyond the area in which a division can effectively employ its combat power. Division deep operations are limited through the use of control measures imposed by the corps and the ranges of the capabilities it controls. During large-scale combat operations, units in the defense likely receive higher CBRN exposure than in the offense. Commanders supported by CBRN staff must ensure that subordinate units can conduct operational decontamination of military personnel and equipment and that they are prepared to execute thorough decontamination based on the priority and type of CBRN exposure. The priority of decontamination support should be established by orders in advance of the operations to expedite action in the event of contamination. These priorities directly support the commander's concept of the operation and scheme of maneuver. Operational decontamination may be conducted in support of committed forces to sustain combat operations. Thorough decontamination sites should be established away from major avenues of approach into the sector and outside the range of the enemy's indirect fire systems.

3-91. The planning, preparation, and execution of uncontaminated (clean) and contaminated (dirty) routes during the defense increase in emphasis due to the higher volume and depth of CBRN exposure to friendly forces. The CBRN staff must assist the operations sections at the division and BCT level with the inclusion of CBRN route graphics on operational overlays. Real-time weather and other environmental factors may change the predictable contamination of routes; therefore, CBRN staff are required to synchronize with each current operations section at echelon.

Decontamination Considerations

3-92. The decision to conduct decontamination should be made only after considering the options that are available based on the type of contamination (persistent, nonpersistent), time available, tactical situation, weather, and readiness of the forces in protective gear. For example, if the agent is a nonpersistent vapor hazard and weather conditions are conducive for weathering of the agent, the practical decision may be to move upwind, conduct unmasking procedures, dispose of potentially contaminated protective clothing, and continue the mission. The risk is not completely mitigated but, based on the tactical situation, the risk of lingering vapor hazards may be lower than the risk of continuing operations with the limitations of MOPP4. Figure 3-10 depicts operational graphics for decontamination sites.

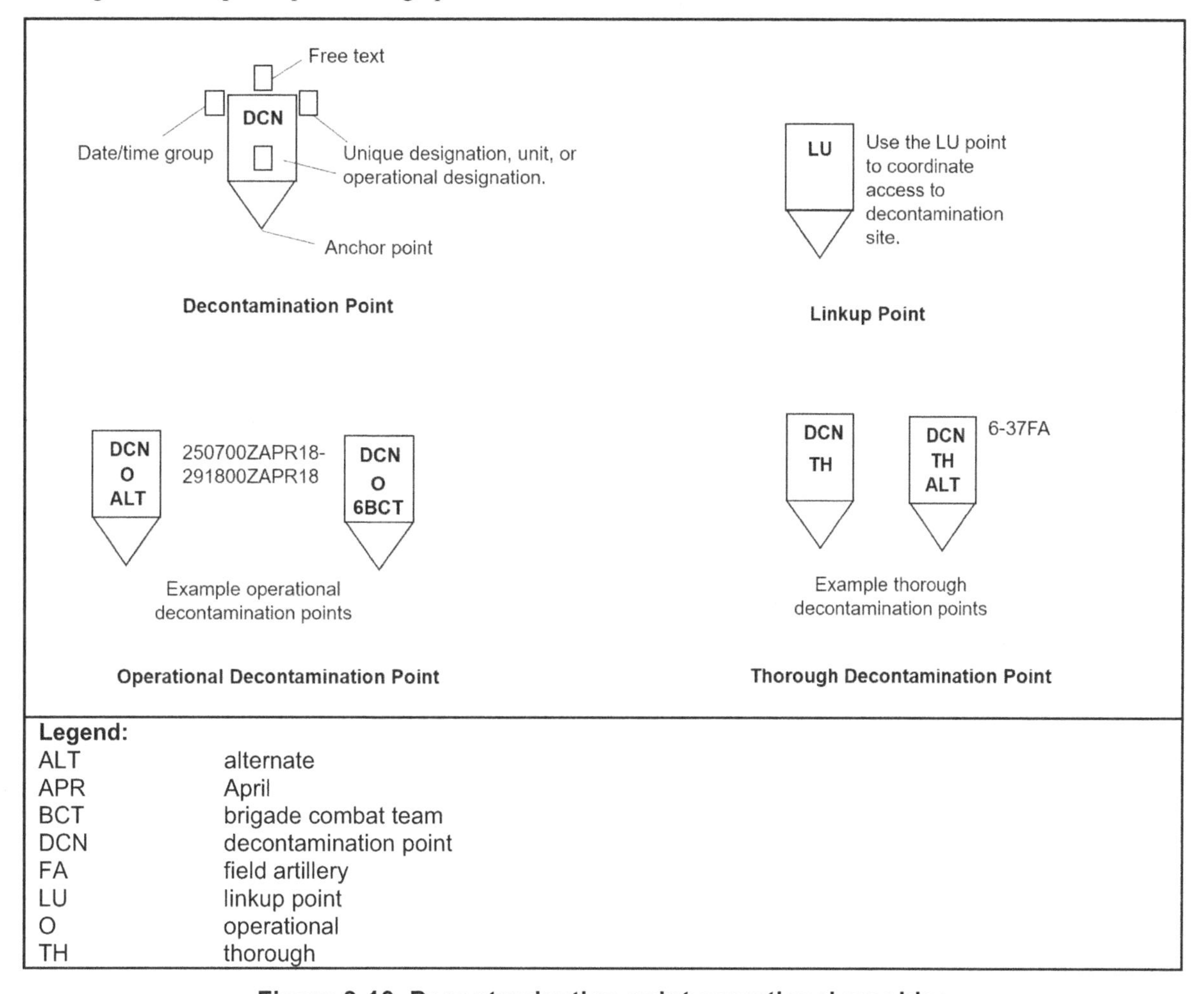

Legend:

ALT	alternate
APR	April
BCT	brigade combat team
DCN	decontamination point
FA	field artillery
LU	linkup point
O	operational
TH	thorough

Figure 3-10. Decontamination point operational graphics

3-93. The consideration for decontamination should be made with the understanding of the following four decontamination principles:

- **Speed.** Personnel should conduct decontamination as soon as possible.
- **Need.** Decontaminate only what is necessary.
- **Priority.** Decontaminate the most essential items first.
- **Limited area.** Decontaminate as far forward as possible to limit the spread to clean areas.

Decontamination Methods

3-94. Contamination occurring on the battlefield is a possibility and could be disastrous; however, planning and training can mitigate the potential effects of such an occurrence. There are four levels of decontamination: immediate, operational, thorough, and clearance.

3-95. The first two levels of decontamination are conducted and supported at the unit level, requiring additional unit training:

- **Immediate.** Immediate decontamination is a lifesaving measure that should be conducted as soon as possible by the individual, buddy, or crew. It includes skin decontamination, personal wipe down, operator wipe down, and spot decontamination. Immediate decontamination should be trained as a battle drill following a CBRN attack and is conducted at the point of contamination.
- **Operational.** Operational decontamination limits the spread of contamination, allows the force to continue operations within the contaminated area, and enables the freedom of maneuver. The tasks include MOPP gear exchange and vehicle wash down. In a CBRN hazard area, the operational decontamination is conducted in a clean area, close to the objective. It is conducted with organic capabilities and the unit trained team.

3-96. The second two levels of decontamination require the technical support of CBRN forces:

- **Thorough.** Thorough decontamination provides a reduction of risk that allows long-term MOPP reduction. The tasks include detailed equipment decontamination and detailed troop decontamination. Outside support from the CBRN unit may be requested, and augmentation is provided to support. Thorough decontamination is resource- and time-intensive, and it is recommended to take place after a unit has completed operations. Detailed troop decontamination is conducted by the unit.
- **Clearance.** Clearance decontamination allows unrestricted transportation, maintenance, and the employment or disposal of equipment. CBRN forces may be called on to advise a commander on support to clearance decontamination. Current United States Army CBRN structures are unable to conduct clearance decontamination to DOD policy standards. Outside agencies are needed to support clearance decontamination.

3-97. Commanders consider operational decontamination when the situation demands continued operations in the contaminated environment. The following are considerations for operational decontamination:

- **When.** Operational decontamination should be planned for within 6 hours of becoming contaminated.
- **Why.** MOPP4 degrades performance, and MOPP gear exchange provides an opportunity for relief. Operational decontamination allows for the removal of contamination so that the unit can conduct unmasking and lower protective posture. Human factor limitations, retrograde, and reconstitution are considerations for operational decontamination.
- **Where.** Decontamination sites are selected via map and site reconnaissance before they are needed. Sites are selected close to a water source, if possible. Designated routes are planned for clean and dirty travel. Preplanning decontamination sites and preparing the site with pre-positioned water can speed up execution when needed.

- **Limiting factors.** The planning time for vehicle wash down is 5-10 minutes per vehicle. The MOPP gear exchange takes approximately 30 minutes for a squad or platoon size element. Limiting factors include water supply, engineer support for drainage, medical support for those working in MOPP and contaminated casualties, individual protective equipment, and decontaminant supply replenishment.
- **Echelon.** A decontamination should be conducted as close as possible to the point of contamination to limit spread; therefore, it is best conducted by organic assets of the contaminated unit. If a larger organization with more vehicles and personnel require operational decontamination and speed is required, the decontamination can be supported by centralized control and support through the brigade and supporting CBRN units.

3-98. If time is available, thorough decontamination should be considered to allow complete reconstitution of the force. Commanders consider thorough decontamination when the situation allows.

HAZARD AWARENESS AND UNDERSTANDING

3-99. During the defense, providing increased hazard awareness and understanding of potential CBRN threats aids the commander's ability to make protection decisions. The CBRNWRS plays a critical role for the early warning of CBRN threats and hazards. The preparation phase of the defense is usually larger than the offense because units are establishing engagement areas (EAs) and battle positions. CBRN staffs must encourage commanders to fully train and rehearse CBRN passive defense plans, which are fed through a proactive risk-based decision-making system. This information improves the evaluation and application of assess/protect/mitigate. Hazard awareness and understanding uses information gained from assess tasks to identify protection and mitigation tasks.

3-100. Through the CBRNWRS, commanders and staffs obtain information on contaminated areas. Hazard awareness and understanding is used to provide the commander with information on areas of residual contamination that affect COA planning for the transition back to offense.

DEFENSIVE PLANNING CONSIDERATIONS

3-101. The purposes of a defense is to deter or defeat enemy offense; gain time; achieve economy of force; retain key terrain; protect the population, critical assets, and infrastructure; and refine intelligence. CBRN staffs, units, teams, and elements provide the necessary assessment, mitigation and protection expertise that enable units in the defense. They recommend effective protective postures, maintain surveillance on CBRN NAIs, recommend or establish fixed decontamination positions and routes, assigned collective protection shelters, and mark CBRN hazard areas to facilitate forward movement and retrograde.

3-102. CBRN staffs support the defense throughout the operations process by applying expertise about friendly and threat capabilities. CBRN expertise is applied to all aspects of the MDMP. Through hazard awareness and understanding, CBRN staffs provide advice to the commander on the implications of potential CBRN hazards and on the impacts of those hazards on COA development. This requires study and analysis to ensure that the right decisions and actions are taken at the right time to get positive outcomes. To prevent the use of WMD, Army forces must develop an understanding of the threat and materials that affect an AO as part of the CWMD mission.

3-103. Planners assess the employment of enemy CBRN capabilities across the operational framework of deep, close, support, and consolidation areas to provide predictive analysis for CBRN threats and hazards that may disrupt units preparing defensive positions and performing terrain-shaping functions. CBRN planners consider the allocation and disposition of decontamination platoons, which can be fixed or mobile throughout the preparation and execution of the defense. The correct placement of these assets limits the spread of contamination, especially for forces retrograding from battle positions. A clear understanding of decontamination control points allows commanders and subordinate units to quickly reconstitute combat power, prioritizing forces assigned to a counterattack.

3-104. Commanders consider mission variables when designating objectives. Special considerations for CBRN environments that the commander and staff consider within the complementary elements when planning defensive operations include—

- **The scheme of maneuver.** In the defense, commanders combine the advantages of fighting from prepared positions, obstacles, planned fires, and counterattacks to isolate and overwhelm selected enemy formations. Commanders must be prepared to rapidly shift the nature and location of their main efforts, repositioning units to mass fires against the attacker to prevent breakthroughs or preserve the integrity of the defense. The defending commander assigns missions, allocates forces, and apportions resources within the construct of decisive, shaping, and sustaining operations and the operational framework. A defensive plan designates axes of advance and routes for the commitment or movement of forces or for the forward or rearward passage of one unit through another. It should identify air movement corridors and other airspace coordination measures to enhance aerial maneuver. The operations process identifies decision points or triggers associated with the initiation of counterattacks, repositioning of forces, commitment of the reserve, execution of situational obstacles, and other actions. All of these actions work to draw advancing enemy formations into the EAs that should eventually defeat enemy offensive operations. CBRN planners must understand the scheme of maneuver at echelon to recommend the correct allocation, mission alignment, and positioning of limited assets according to the CBRN protect, assess, and mitigate functions.
- **Planning operations.** A corps or division commander employs fires to neutralize, suppress, or destroy enemy forces. Fires disrupt an enemy's ability to execute a preferred COA. Corps and division staffs assist their commanders to integrate indirect fires, electronic attacks, aviation maneuver, and joint fires into the defensive plan. This allows the defending commander to degrade the enemy before entering the main battle area (MBA). Part of the targeting process is deciding which enemy systems and capabilities to attack. The CBRN staff uses assessments from the planning process to recommend adjustments to the attack guidance matrix to mitigate the enemy's CBRN capability before employment. In the defense, expect a higher volume of enemy fires, some of which could be mixed with CBRN munitions. When planning fires, the commander must consider intelligence about the enemy's CBRN capabilities and the impact of hazard areas created from counterforce actions. CBRN planners need to advise the commander on the employment of CBRN reconnaissance assets in deep operations. Planners must consider the protection of supporting assets to counter enemy fires.
- **Reconnaissance and security operations.** Commanders use reconnaissance and security operations to confuse the enemy about the location of friendly EAs. These operations also prevent enemy observation of friendly defensive preparations which limits their ability to use observed fires on those positions. The use of active and passive counter-reconnaissance efforts prevent the enemy from determining the precise location and strength of the defense. The proper size, positioning, and timing of forward security forces can cause an attacking enemy to deploy prematurely. These forces can either assume concealed hide-positions or conduct a rearward passage of lines before the enemy's main attack enters EAs. Aggressive security operations in the close and support areas deny enemy reconnaissance and action from special purpose and irregular forces. CBRN planners support reconnaissance and security operations by advising the commander on the proper alignment of CBRN assets along the protect, assess, and mitigate functions. This may include CBRN R&S assets assigned to larger reconnaissance and security operations or directed to answer specific CBRN PIR within the deep and close areas. Furthermore, CBRN R&S elements can augment these operations to provide early warning of CBRN hazards. All forces make use of the CBRN warning and reporting system to provide early warning of CBRN hazards and allow immediate protection to friendly forces operating in the area. CBRN passive defense measures are taken to protect friendly forces, installations, routes, and actions within a specific area. Security operations (screen, guard, and cover) provide early warning and protection to the maneuver force. The proper allocation of decontamination assets can further assist commander with reconstituting security forces compromised by CBRN exposure. Again, CBRN assets directly support these operations with critical enablers. More information about CBRN reconnaissance, including route, area, and zone, can be found in ATP 3-11.37.

- **Reserve or strike force operations.** The defense plan retains a reserve regardless of the form of the defense. Mobile defenses at division levels and higher call for a robust strike force, which can be as large as one-third of the commander's available combat power but is normally not smaller than a BCT size element for a corps. A tactical reserve or strike force is an uncommitted force that is available for commitment at the decisive moment. Commanders designate planning priorities with supporting, on-order graphic control measures for commitment of the reserve. Commanders plan the movement of the reserve from its AA to probable commitment locations to deconflict with other movements, including sustainment convoys, artillery repositioning, and other maneuver forces within their AOs. Upon commitment, the reserve or strike force is supported with additional assets (artillery, close air support, attack aviation, electronic warfare, CBRN defense, sustainment). CBRN planners must prioritize their protect, assess, and mitigate functions to support the reserve or strike force and enable freedom of maneuver.
- **Retrograde in the defense.** The most common retrograde in the defense is the rearward passage of lines for a forward security force. The area defense or mobile defense can include other retrograde tasks (delay, withdrawal, retirement). In certain circumstances, a retrograde may be forced or voluntary. In all cases, the higher echelon commander must approve the retrograde because improper synchronization could place multiple forces at a higher vulnerability to enemy attacks. The complexity and fluidity of retrograde require centralized planning and decentralized execution. CBRN planners support retrograde by advising commanders about the likelihood of enemy CBRN employment to fix friendly forces and how to best mitigate contamination as these elements move rearward.
- **Support area operations.** The success of the defense depends heavily on protecting the corps and division support area from enemy attack. Planning for area security is a critical role for the CBRN force because these areas have the preponderance of critical assets and infrastructure. The wide range of threats to key nodes in the support area include, but are not limited to, individual saboteurs, airborne or air assault insertions, long-range artillery and missile strikes, and ground maneuver penetrations. Attacks against friendly forces in the support area, especially from small units, may precede the onset of large-scale combat operations and be indistinguishable from terrorist acts. The commander defines responsibilities for the security of units within the support area. The MEB commander's AO is a support area. The MEB commander is responsible for area security within the AO and sets protection standards. Corps and division commanders carefully balance available assets against requirements to determine acceptable risks for the support area during the defense. The dispersion of mission command nodes and support capabilities prevent the enemy from identifying lucrative targets. The MEB uses its military police and other assigned security forces to provide area security. CBRN planners further advise corps and division commanders about additional considerations to mitigate risks to the support area by aligning priorities and assets that support the concept of operations for the defense. They consider the protection of, and recommend alternatives for, ground lines of communications that support movement to the close area and may be at risk to CBRN contamination.

CBRN SUPPORT TO DEFENSIVE OPERATIONS

3-105. There are three defensive operations that operational and tactical commanders can utilize when assuming or transitioning to the defense: area defense, mobile defense, and retrograde. Commanders make decisions on the appropriate operation based on the current conditions of the OE and the strength or will of the enemy. Although these operations can have static and mobile elements, they are significantly different concepts and pose unique challenges in planning and execution. Using only static defensive elements puts friendly forces at a greater disadvantage, especially when the enemy can outrange the defender's fire support systems or when it has a larger capacity to engage with indirect fires. All three operations use terrain, depth, and mutual supporting fires as force multipliers. Typically, divisions are the lowest level of command that can successfully execute a mobile defense because of the combat power, planning, and resources required to execute. The same consideration is applied to complex retrograde tasks that involve a delay, withdrawal, or retirement of multiple battalions.

3-106. There are three subordinate forms of defense: defense of a linear obstacle, perimeter defense, and reverse-slope defense. Each of these has special purposes and different planning considerations. The mobile aspects of the defense can utilize appropriate forms of maneuver, such as frontal attack, flank attack, envelopment, and turning movement. Other mobile aspects involve simple forms of retrograde, such as the rearward passage of lines from forward security elements. There are seven characteristics of the defense that commanders and staffs consider in planning and execution: disruption, flexibility, maneuver, massing effects, operations in depth, preparation, and security.

Note. See FM 3-90-1 for additional information on defensive tasks and subordinate forms and characteristics of the defense.

3-107. Commanders organize friendly forces for defensive operations based on the echelon engaged in defensive tasks, how the terrain allows major avenues of approach, the attacking enemy's size and capabilities, and the type of defensive task applied to achieve the intent of the operation. Corps and division commanders organize their defense in depth along a contiguous or noncontiguous framework for large-scale combat operations with deep, close, support, and consolidation areas. The decision for a noncontiguous framework is weighted by the size and type of terrain and the available avenues of approach. Regardless of the framework, the preponderance of ground combat power is focused on defeating the enemy in a main battle area aligned to key terrain inside the close area. The general organization for the defense includes information collection, security, MBA, reserve, and sustaining forces.

3-108. Ideally, committed divisions and BCTs have enough depth to provide security through the employment of a covering or a guard force. BCT commanders typically assign their reconnaissance squadron as a forward security force to maintain contact with enemy reconnaissance elements to deny observation of defensive preparations and to prevent disruption through observed fires and limited forms of contact. A sustaining force with imbedded fires in the support area is critical to any defense because of the resources required for defensive preparation and the necessity to shape the deep fight with long-range fires and combat aviation. The subordinate commander who controls the support area (typically the MEB for divisions) has a tactical combat force that provides area security and can react to a level III threat, which involves enemy armor organized at company team size or larger. Corps commanders normally assign a BCT as a tactical combat force if all subordinate forces are involved in a defense. Commanders always maintain a reserve, regardless of the defensive task. A mobile defense requires the commander to maintain a strike force, which can take up to one-third of the available combat power.

3-109. Based on the size, composition, and direction of the enemy attack, commanders select the best location to defeat or destroy enemy assault exploitation forces at the point of penetration. The reserve may counterattack into the enemy flank, or it may establish a defensive position in depth to defeat or block further enemy advances. The corps and division staffs establish control measures for counterattacks by developing a supporting scheme of fires with target areas of interests.

3-110. The defense at the corps level typically requires the involvement of a CBRN brigade headquarters that has an allocation of at least one CBRN battalion headquarters to the division conducting the decisive operations. Because of a higher likelihood of enemy CBRN employment and a higher expected volume of fire, these headquarters augment planning and recommend the allocation of their companies in support of CBRN passive defense measures.

3-111. CBRN R&S assets support the depth of the defense to provide detection and marking along key terrain and essential movement corridors required to successfully execute the concept of operations and the scheme of maneuver. These assets are organized, allocated, and assigned at the platoon level and focus on information collection in support of CBRN PIRs for commanders at the BCT level and higher. CBRN staffs also recommend employing mounted CBRN R&S at echelons above brigade in the defense to monitor key ground lines of communication that support main supply routes essential for defensive preparations. CBRN R&S elements and platoons are employed with security forces to provide early warning of CBRN attacks. CBRN R&S (mounted and dismounted) and bio-detection units can support critical assets and sites by providing versatile capabilities to inform the commander of any detection of CBRN employment and likely hazard areas. CBRN mounted and dismounted platoons, teams, and elements mark the extent of contamination and bypass routes to afford combat power to enter and move throughout the defensive area.

3-112. Additional CBRN decontamination units are task-organized with divisions and BCTs and are pre-positioned to support the reconstitution of essential combat power contaminated during CBRN attacks on likely targets. Within the scope of a mobile defense, CBRN decontamination assets should be weighted toward ensuring that the strike force maintains freedom of action. For an area defense, decontamination assets should be positioned to maximize support to artillery assets; they are essential to mission accomplishment and to the ability to wither enemy offensive capability. CBRN staffs must also plan and prepare for contaminated equipment to retrograde along designated planned contaminated routes for decontamination support. Contaminated units and teams must be prepared to conduct MOPP gear exchange and detailed troop decontamination. CBRN information should be reported for situational awareness and situational understanding.

3-113. CBRN staffs recommend employing mounted CBRN R&S at echelons above brigade in the defense to monitor key lines of communication that support key routes for all classes of supply to build and improve defenses and sustain the force. CBRN R&S elements and platoons are employed with scouts to monitor the enemies that are probing and posturing for the attack. CBRN mounted and dismounted R&S detect and identify enemy CBRN employment and hazard areas. CBRN mounted and dismounted platoons, teams, and elements mark the extent of contamination and bypass routes to afford combat power to enter and move throughout the defense area.

3-114. CBRN decontamination teams or elements are pre-positioned and prepared for contaminated equipment to retrograde along planned contaminated routes for decontamination support. Contaminated units and teams are prepared to conduct MOPP gear exchange and detailed troop decontamination when necessary. CBRN information is reported for situational awareness and situational understanding. CBRN decontamination units are task-organized to maximize the decontamination of equipment to enable maneuver, fires, and sustainment personnel to continue the mission.

Support to Defensive Operations (Mobile Defense)

3-115. A mobile defense focuses on destroying the attacking force by permitting the enemy to advance into a position that exposes the enemy to counterattack and envelopment. The commander retains the majority of available combat power in a striking force for the decisive operation (a major counterattack). The commander commits the minimum possible combat power to the fixing force that conducts shaping operations to control the depth and breadth of the enemy's advance. The fixing force also retains the terrain required to conduct the striking force decisive counterattack. On the other hand, the area defense focuses on retaining terrain by absorbing the enemy into an interlocked series of positions, where the enemy is destroyed largely by fires.

3-116. Army commanders organize the main body of a mobile defense into two principal groups—the fixing force and the striking force. Units smaller than a division do not normally conduct a mobile defense because of their limited capabilities to fight multiple engagements. In the mobile defense, reconnaissance and security, reserve, and sustaining forces accomplish the same tasks as in an area defense. The fixing force has the minimum combat power needed to turn, block, and delay the attacking enemy. This usually means that the defending force must allocate a majority of its countermobility assets to the fixing force to shape the enemy penetration or contain the enemy's advance. Typically, the striking force may consist of one-half to two-thirds of the defender's combat power. It decisively engages the enemy as attacking forces become exposed in their attempts to overcome the friendly fixing force. Resourcing a reserve in a mobile defense is difficult and requires commanders to assume risk. Commanders generally use the reserve to support the fixing force. However, if the reserve is available to the striking force, it exploits the success. The commander completes required adjustments in task organization before committing subordinate units to combat.

3-117. Several control measures assist the commander in synchronizing mobile defense operations. These control measures include designating the AOs of the fixing and striking forces with their associated boundaries, battle positions, and phase lines (PLs). The defending commander designates a line of departure or line of contact as part of the graphic control measures for the striking force. The commander may designate an axis of advance for the striking force with attack by fire or support by fire positions. EAs, target reference points, target areas of interest, and final protective fires allow for further coordination and synchronization of direct fires systems with indirect fires systems. The commander designates NAIs to focus the efforts of information collection assets. This allows the commander to determine the enemy's chosen COA. The commander designates checkpoints, contact points, passage points, and passage lanes for use by reconnaissance and surveillance assets, security units, and the striking force.

3-118. Figure 3-11 provides a COA sketch for an Army division organized to conduct a mobile defense. This sketch further illustrates the integration of CBRN forces to support CBRN passive defense measures and the requirements for the assess, protect, and mitigate functions. The division is organized with an organic armored, Stryker, and infantry brigade combat team (IBCT) and with division artillery, attack aviation, and support brigades. Several units are configured with specific task organizations to allow a mixture of combat power and appropriate enabling assets. An MEB, battlefield surveillance brigade, and CBRN battalion are mission-aligned to the division in support of the mobile defense. The division is using its battlefield surveillance brigade as a security and reconnaissance force. A task-organized Stryker BCT (SBCT) and a task-organized IBCT are fixing forces in this mobile defense and are in prepared battle positions with enhanced EAs that have specific obstacle effects. The fixing force is using terrain canalized by mountains to prevent penetration beyond forward battle positions. A task-organized ABCT is the division's striking force and is further augmented with attack aviation and division fires. Finally, the division has one combined arms battalion from the ABCT as a reserve, which is pre-positioned near the IBCT with the division tactical command post.

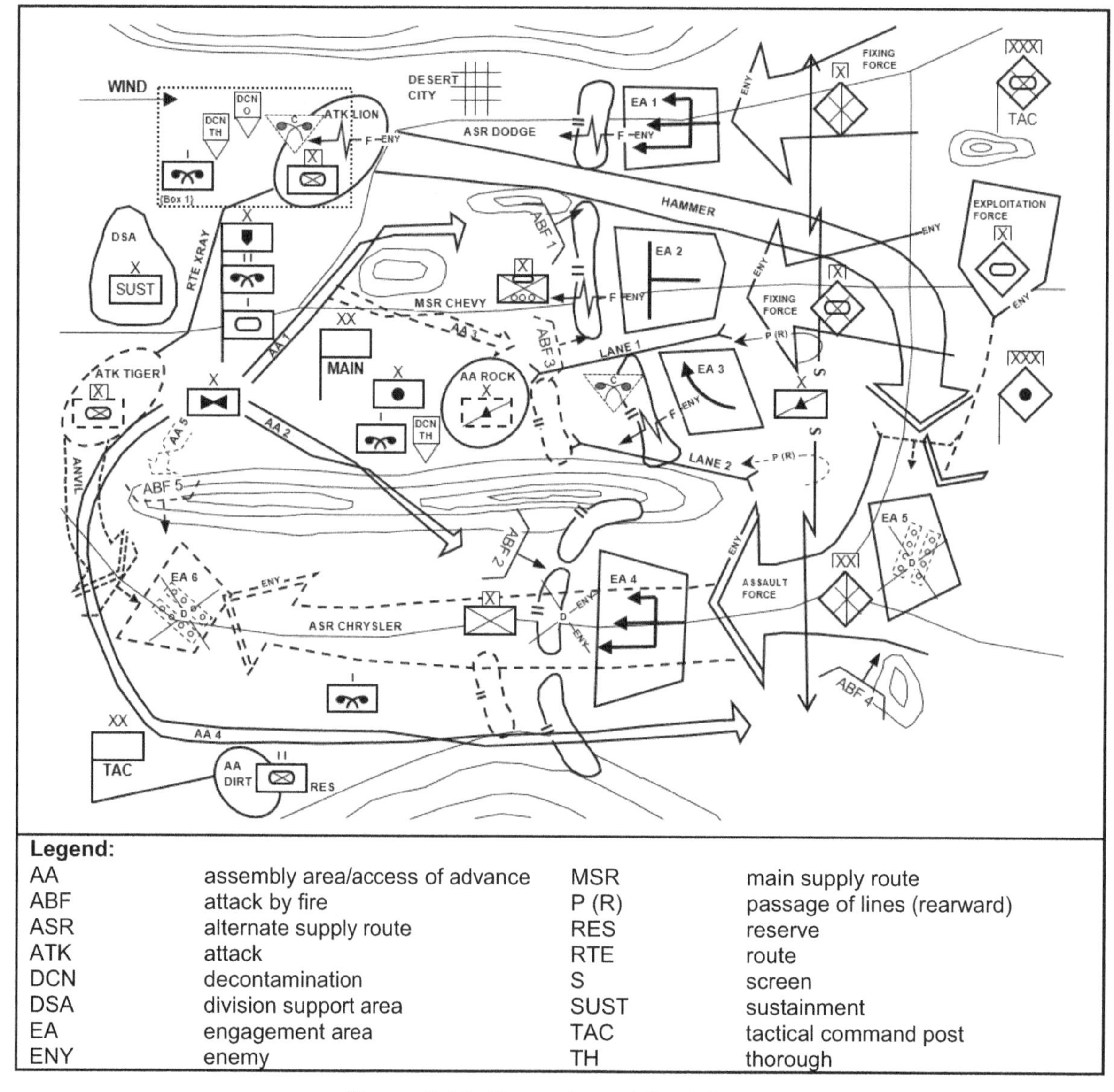

Figure 3-11. Example mobile defense

3-119. The enemy in figure 3-11 represents a motorized-infantry division tactical group augmented with a tank brigade tactical group and mechanized infantry brigade tactical group. The addition of an enemy corps tactical command post and integrated fires command is meant to replicate the main effort of an operational-strategic command in the attack against an opposing friendly corps in the defense. The enemy is using an integrated attack with enabling and action forces to defeat its adversary. The enemy commander is using a task-organized motorized brigade tactical group and a mechanized infantry brigade tactical group as fixing forces against its adversary's most heavily defended area. A task-organized motorized-infantry division tactical group is the assault force for a tank brigade tactical group, which is the exploitation force. The operational-strategic command has task-organized its combat power across the division tactical group and brigade tactical groups to create a mixture of maneuver and support assets. It intendeds to penetrate along canalized terrain, at the bottom of the sketch, where it expects its adversary to defend with less combat power. The objective of the exploitation force is to penetrate into its adversary's support area to cause its defeat and allow for a follow-on division tactical group to continue the attack.

3-120. The COA depicted in figure 3-11 employs an ABCT strike force as the decisive operation along axis of advance HAMMER to destroy the enemy exploitation force in EA five (EA 5) to prevent enemy penetration of the division's main defensive belt. The battlefield surveillance brigade, the division's security force, initially conducts a forward screen to allow friendly defensive preparation. On order, the battlefield surveillance brigade conducts a rearward passage of lines with the SBCT along passage lane one (LANE 1) to draw the enemy main attack into the division's primary EAs. The attack aviation brigade supports the division's fixing and striking force along multiple air axes of advance with attack-by-fire positions that are oriented at EAs. The division commander has created alternate employment options for its strike force. The COA also depicts the movement of the ABCT strike force along an alternate axis of advance ANVIL to destroy the exploitation force in EA six, which is well inside the close area.

3-121. The enemy's employment of CBRN is shown in figure 3-11 with a suspected nonpersistant chemical attack on the SBCT to disrupt friendly combat power in battle positions to enable its assault and exploitation force attack on the IBCT. The enemy commander utilizes its forward reconnaissance and irregular assets in the nearby urban area (DESERT CITY) to further pinpoint locations of the friendly strike force and of command, fires, aviation, and logistics nodes. The COA in figure 3-11 displays an enemy persistent chemical attack on the strike force.

3-122. The CBRN battalion aligned to the defending division has additional decontamination capabilities due to the likelihood of enemy CBRN employment. Hazard response companies have been allocated to support the IBCT and SBCT. An additional hazard response company is pre-positioned at a brigade-level decontamination point for thorough decontamination with the ABCT strike force's designated decontamination team to support the rapid reconstitution of the strike force if contaminated.

3-123. In a situation for which it is expected that the enemy will employ CBRN agents, operational and thorough decontamination points may be established in advance to expedite the ability of the force to prioritize and restore combat power. Figure 3-12, page 3-32, shows an expansion of Box 1 from figure 3-11, in which the BEB and CBRN hazard response company have established a linkup point where units are sorted for operational or thorough decontamination. Vehicles that are essential combat power to the brigade mission and that may continue to fight in a contaminated area are processed through the operational decontamination point. The thorough decontamination point is used for units that are mission-essential but must operate with minimum degradation from CBRN contamination or risk to troops. Two decontamination points facilitate the increased throughput of rapid reconstitution of the strike force combat power.

3-124. If the armored brigade becomes contaminated (and depending on the assessment of the principles of decontamination [speed, need, priority, and limited area]), the commander has the following options:

- Send contaminated elements of the brigade for operational decontamination if the brigade needs to continue the fight in the contaminated area but wishes to remove gross contamination before moving forward.
- Send contaminated elements that are out of the fight to the thorough decontamination point to reduce the contaminated unit MOPP level and to fully reconstitute it.
- Continue operations, fighting dirty to the extent that the commander can assume risk.

3-125. The required engineering support and sustainment that must occur to simultaneously operate an operational and thorough decontamination site are critical to mission success. The established sites have drainage sumps dug with engineer support and pre-positioned water available. Sustainment units must actively push water, decontamination solution, individual protective equipment sets, medical CBRN defense materiel, and replacement parts (especially air filters) to decontamination sites and AAs, using only designated clean routes to prevent the spread of contamination. Figure 3-12 also shows a contaminated (dirty) casualty collection point from which contaminated casualties can be assessed at marshalling areas and transferred to supporting medical units.

Note. See ATP 4-46 for more information on mortuary affairs.

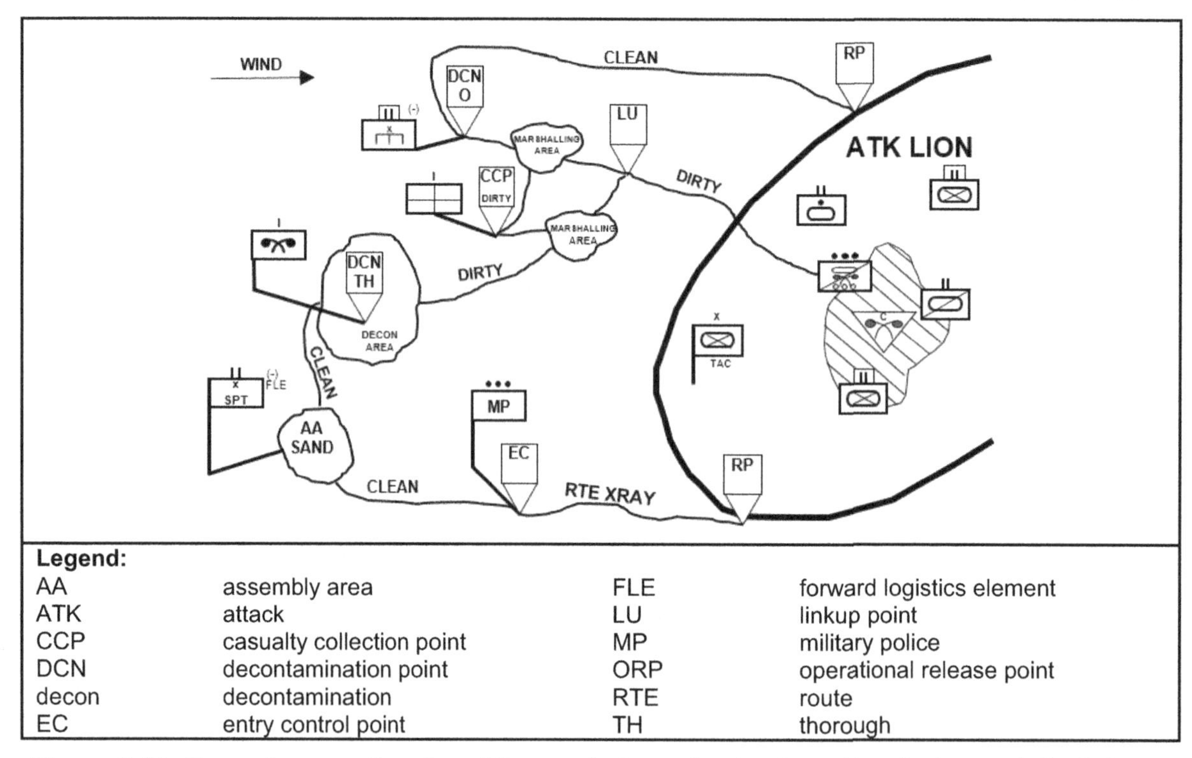

Legend:

AA	assembly area	FLE	forward logistics element
ATK	attack	LU	linkup point
CCP	casualty collection point	MP	military police
DCN	decontamination point	ORP	operational release point
decon	decontamination	RTE	route
EC	entry control point	TH	thorough

Figure 3-12. Example operational and thorough decontamination sites during mobile defense

CONSOLIDATE GAINS

3-126. Commanders have few resources to devote to the consolidation of gains during tactical operations focused on the defense. The primary consolidation of gains activities during operations focused on the conduct of defensive tasks ensures that the future conduct of consolidation of gains tasks is adequately addressed in sequel and branch plans to the current operations order.

3-127. Consolidation of gains activities in the defense may include marking contaminated areas, conducting decontamination, and assisting essential services. CBRN biological surveillance and area support companies should retain operational agility to surge from corps or division consolidations areas to support other CBRN unit mitigation tasks during the defense, particularly during large-scale CBRN strikes that overwhelm hazard response company capabilities.

SECTION III—CBRN CAPABILITIES IN STABILITY OPERATIONS

OVERVIEW

3-128. The body of security strategy that shapes the conduct of operations characterized by stability operations includes the National Security Strategy, National Defense Strategy, and National Military Strategy and related strategies, including the National Strategy for Countering Weapons of Mass Destruction. Together with national policy, strategy provides the broad direction necessary to conduct operations to support national interests.

STABILITY OPERATIONS

3-129. *Stability operation* is an operation conducted outside the United States in coordination with other instruments of national power to establish a secure environment while providing essential governmental services, emergency infrastructure reconstruction, and humanitarian relief (ADP 3-0). As the Army transitions out of conflict, the following principles of stability tasks lay the foundation for long-term stability: conflict transformation, unity of effort and unity of purpose, legitimacy and host-nation ownership, and building partner capacity.

3-130. Stability mechanisms, applicable across the competition continuum, are the primary method by which the joint force affects the human dimension. Operations conducted during the stability phase may be conducted unilaterally and include up to the joint force or the whole of government and unified action partners. Operations are conducted concurrently and address a COCOM's theater campaign plan lines of effort that provide a realistic appraisal of relevant partner relationships, which allows the commander and staff to derive a range of feasible, productive military options that lead to sustainable and acceptable outcomes.

CONSOLIDATE GAINS

3-131. *Consolidate gains* are activities to make enduring any temporary operational success and set the conditions for a stable environment allowing for a transition of control to legitimate authorities (ADP 3-0). The consolidation of gains is an integral and continuous part of offense, defense, and stability. Commanders continuously consider the synchronization, integration, and organization of protection capabilities necessary to consolidate gains and achieve the desired end state. When consolidating gains, establishing and sustaining security is first priority.

3-132. The consolidation of gains occurs in portions of an AO for which large-scale combat operations are no longer occurring. Consolidation of gains activities consist of security and stability tasks and will likely involve combat operations against bypassed enemy forces and remnants of defeated units. Therefore, units may initially conduct only minimal essential stability tasks and then transition into a more deliberate execution of stability tasks as security improves.

3-133. Operations to consolidate gains require combined arms capabilities and the ability to employ fires and manage airspace, but at a smaller scale than large-scale combat operations. Units in the close area involved in close combat do not conduct consolidation of gains activities. Consolidation of gains activities are conducted by a separate maneuver force in the designated corps or division consolidation areas.

3-134. Commanders must reduce postconflict or postcrisis turmoil and help stabilize a situation. Gains may include the establishment of public security by using friendly forces to transition, the performance of humanitarian assistance, and the restoration of key infrastructure. The MEB supports consolidation of gains and is tailored with enabling forces as needed to support the situation. The effectiveness and success of the MEB depend on the synergy that is leveraged from integrating the contributions from enabling units, such as CBRN.

Note. See FM 3-81 for more information on MEB.

3-135. CBRN forces conduct tasks of consolidating gains in support of stability. Examples of the tasks conducted by CBRN forces include, but are not limited to, expertise support, humanitarian assistance, host-nation leadership engagement, security cooperation engagement, and foreign internal defense. To deter a theater threat's potential employment of CBRN capabilities, the geographic combatant commander coordinates the conduct of security cooperation engagements with adjacent neutral and ally countries to understand mutual capabilities, synchronize effects, and gain efficiencies within the coalition. CBRN forces, with other governmental agencies, increase the effectiveness and efficiency of neutral and ally internal infrastructure assessments, demilitarization, and rebuilding efforts. Consolidate gains activities include CBRN response tasks that restore and protect critical infrastructure, equipment, and personnel.

3-136. Army forces conduct continuous reconnaissance and, if necessary, gain or maintain contact with the enemy to defeat or preempt enemy actions and retain the initiative. Consolidating gains may include actions required to defeat isolated or bypassed threat forces to increase area security and protect lines of communication. The discovery of WMD sites, TIMs, or CBRN contamination during decisive action tasks presents unique challenges for consolidating gains. CBRN forces must confirm or deny the presence of CBRN hazards. Maneuver units must achieve a minimum level of control to create conditions for success in dealing with these materials. Commanders address the decontamination, disposal, and destruction of CBRN materiel and WMD.

3-137. Tasks within CBRN functions support the consolidation of gains. Activities within the assess function (such as information collection and exploiting WMD sites) feed CCIR and allow the commander to exploit further tactical gains. The assess function also feeds into hazard awareness and understanding, increasing the commander's situational understanding and allowing for the refinement of follow-on operations. The CBRN function of protect includes tasks of CBRN defense, such as the protection of forces from CBRN hazards as gains are consolidated. Tasks within the mitigate function respond to CBRN effects to negate hazard effects.

Threat Overview

3-138. The threat overview describes how threat forces organize again U.S. forces during stability. This allows CBRN forces to understand enemy potential courses of action.

3-139. While U.S. forces seek to establish stability, asymmetric threats still have the potential to use CBRN hazards to gain advantage and to use them as a major destabilizing force. Nonstate actors will actively pursue and use CBRN materials to overcome force overmatch and create conditions favorable to their cause. It is highly likely that CBRN usage from asymmetric threats will tie directly into a sophisticated information or disinformation campaign. As a destabilizing force, asymmetric threats may utilize CBRN materials to provoke sectarian violence, conduct mass atrocities, or incite public panic. Threats may attempt to destabilize the perception of civil security by attacking transportation nodes and water sources and assassinating key leaders and government officials as part of their information warfare campaigns. Critical to the threat's use of CBRN materials is maintaining technically competent individuals who can identify, cultivate, and create delivery mechanisms for WMDs or hazardous substances. Often, this requires small-scale chemical or biological production facilities or clandestine labs that operate using dual-use equipment. Therefore, it is imperative for CBRN assets and staffs to quickly identify internal and external threats that have the technical capacity for using CBRN. Within hybrid warfare, state actors may support insurgencies, criminal elements, contractors, or special-purpose forces that have technical expertise or facilities to enable asymmetric CBRN usage.

Protect

3-140. During stability and as done in offense and defense, CBRN enablers continue to provide protection to forces. CBRN enablers continue to conduct reconnaissance and surveillance as required, provide assessments on protective postures, and maintain a warning and reporting system. CBRN forces may be required to protect the local population from CBRN hazard areas, identifying and marking areas in support of the area security plan.

MITIGATE

3-141. The mitigate function helps commanders to execute tasks to reestablish a safe and secure environment, provide humanitarian relief, and minimize human casualties when CBRN incidents exceed host-nation capabilities. CBRN response activities may be requested by the Department of State and directed by the Secretary of Defense, DOD, as a part of international chemical, biological, radiological, and nuclear response (ICBRN-R). ICBRN-R applies to international incidents involving the deliberate or inadvertent release of CBRN materials. If the affected nation requires assistance for handling a CBRN hazard, U.S. assets may augment the affected nation's assets to restore stable conditions. Major functions performed include safeguarding lives, preserving health and safety, securing and eliminating the hazard, protecting property, and preventing further damage to the environment. More about ICBRN-R can be found in appendix C and JP 3-41.

3-142. During the course of large-scale combat operations, sensitive sites may be discovered or direct action may create CBRN hazard incidents that are beyond the capabilities of the host-nation resources and require the support of CBRN forces to stabilize the situation. Sensitive sites are known or suspected of involvement in enemy research, production, storage, or past or future employment of CBRN weapons. Either situation may require resources of CBRN and other technical enablers to mitigate the hazards. Responding to CBRN effects is a key component of efforts to reconstitute the force to prepare for follow-on missions. CBRN tasks that support responding to CBRN effects include—

- **Mitigating contamination.** Mitigating contamination can be achieved by controlling contamination; limiting the vulnerability of forces to CBRN and TIM; and avoiding, containing, and controlling exposure. The decontamination of personnel, equipment, and facilities depends on the time and resources available.
- **Controlling, defeating, disabling, and/or disposing of WMD.** WMD defeat is undertaken to systematically destroy WMD materials and related capabilities. Support to CWMD operations must initially focus on the immediate tasks of control (seizing and securing sites and preventing the looting or capture of WMD and related materials) and the disablement or destruction of weapons, materials, agents, and delivery systems that pose an immediate or direct threat to forces and the civilian population. Expedient tactical WMD defeat, disablement, and/or disposal may be required to ensure the safety of troops, secure the freedom of action for combat operations, or protect noncombatants. EOD is a key component of WMD defeat, disablement, and/or disposal. EOD provides significant expertise by supporting technical intelligence collection and exploitation, providing guidance on protective measures, conducting render safe, assisting in destruction of transfer activities, and supporting monitoring and redirection efforts.

Note. See ATP 4-32 for more information on EOD operations.

3-143. During stability, the CBRNE command engages in the assessment, protection, and mitigation of CBRNE hazards. Nuclear facility disablement, EOD final disposition, chemical and biological laboratory operations, CBRNE modeling, and reachback are some of the tasks conducted. A CBRNE command headquarters can also advise partner nations or host-nation authorities and build partner capacity in establishing and training partner CBRN and EOD forces to provide for civil security.

ASSESS

3-144. During stability, the CBRN staff continuously conducts assessments of CBRN threats and vulnerabilities. Stability activities support the prevention of continued conflict. CBRN enablers support the prevention of potential CBRN incidents through inspections and monitoring. CBRN reconnaissance elements assess and characterize sites to confirm or deny potential follow-on mitigation efforts.

CBRN ROLE IN SUPPORT OF AREA SECURITY OPERATIONS

3-145. By definition, area security operations focus on the protected force, installation, route, or area. As a part of area security, the consolidation of gains force may have to identify and destroy remaining pockets of enemy forces. Search and attack is the technique used to destroy the enemy and protect the force.

3-146. Figure 3-13 depicts a corps noncontiguous area. Within the area, U.S. forces conduct area security while remnants of the enemy force still exist in the area. A joint task force is formed to cordon and search a suspected WMD site in the city used to store chemical weapons. The joint task force must secure the site on OBJ TANGO and conduct sensitive site exploitation of the chemical storage facility to ensure that CBRN hazards do not fall into enemy hands. The joint task force elimination headquarters maintains control of the area operations, including OBJ TANGO. Three IBCTs have secured the surrounding terrain around OBJ TANGO to prevent enemy maneuver during sensitive site exploitation.

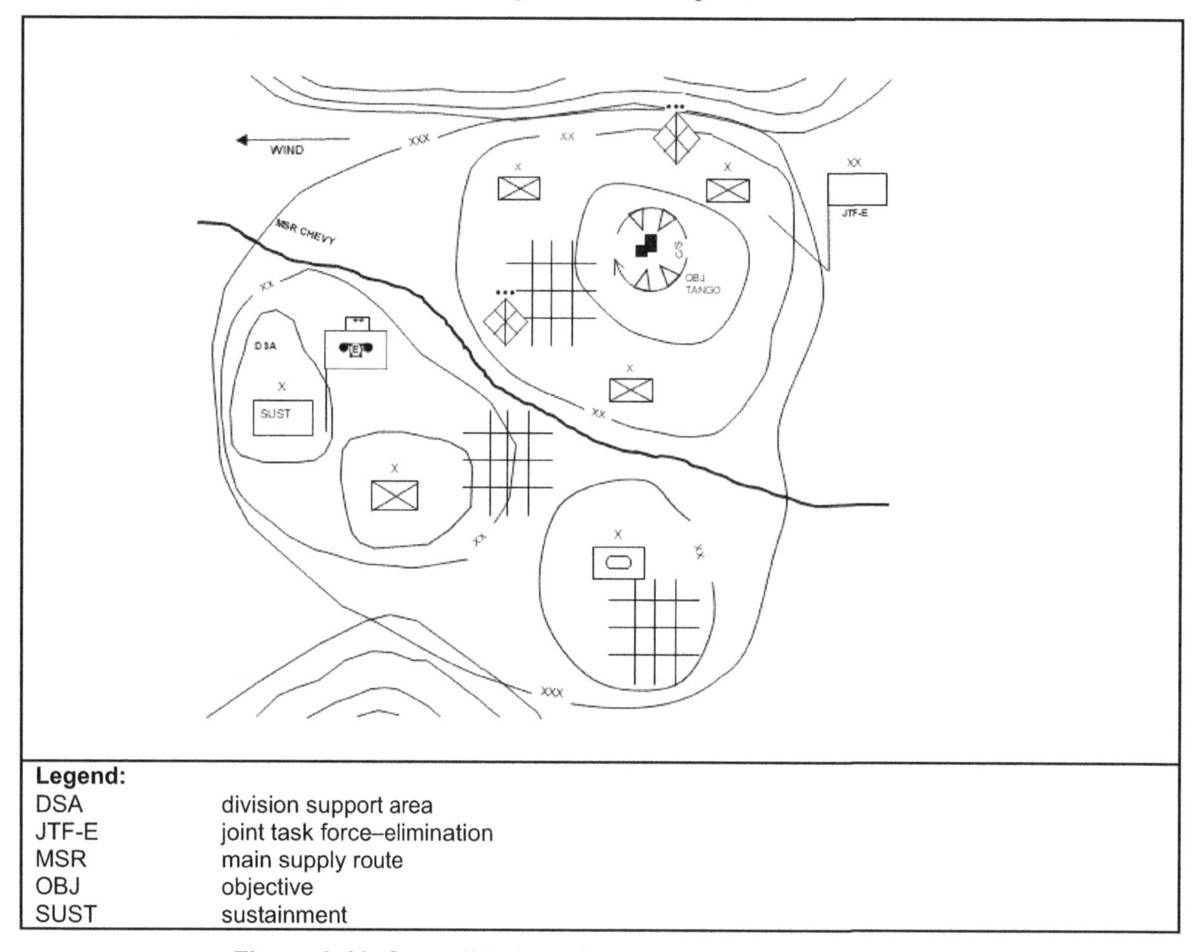

Legend:

DSA	division support area
JTF-E	joint task force–elimination
MSR	main supply route
OBJ	objective
SUST	sustainment

Figure 3-13. Consolidation of gains in a noncontiguous area

3-147. Figure 3-14 depicts the integration of CBRN enablers for the cordon and search of the chemical storage facility. An infantry battalion task force attached with a CBRNE company, a hazard response company, and a medical support company with role 1 capability are depicted. Planners should consider increased security elements when the tactical situation dictates.

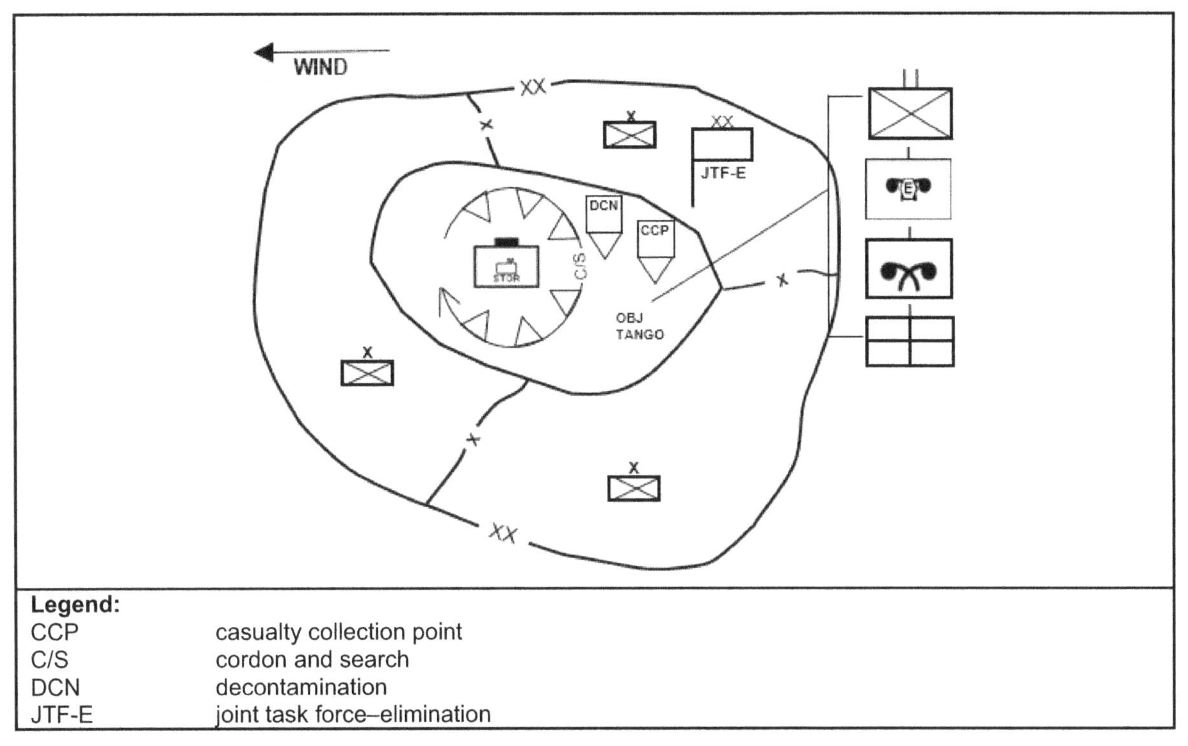

Legend:

CCP	casualty collection point
C/S	cordon and search
DCN	decontamination
JTF-E	joint task force–elimination

Figure 3-14. Cordon and search of a sensitive site

CBRN ROLE IN SUPPORT TO STABILITY OPERATIONS

3-148. Stability is achieved through stability mechanisms (compel, control, influence, and support). Stability mechanisms are the primary method through which friendly forces affect civilians to attain conditions that support establishing a lasting, stable peace. CBRN staffs, teams, and units provide protection and mitigate CBRN hazards to enable stability operations tasks (civil security, civil control, restoration of essential services, governance, economic and infrastructure development, and security cooperation). CBRN staffs assess TIM sites and suspected chemical biological manufacturing plants that may have been damaged during conflict. Staffs understand that they must work in and through the military and civilian control to provide capability at the point of need. CBRN capability provides mounted and dismounted R&S at damaged facilities that may have been used as research and development or for the manufacturing of chemical or biological weapons. CBRN staffs conduct battle damage assessments of TIM sites that may affect the health and safety of the civilian populace. CBRN decontamination platoons provide equipment decontamination to critical police, ambulances, fire trucks, engineer construction equipment, and infrastructure to enable civil security and civil control. CBRNE response teams and hazardous assessment platoons conduct the exploitation and destruction of chemical and biological munitions. Nuclear disablement teams conduct assessments of nuclear power plants.

Note. See ATP 3-90.40 for more information on employment considerations of CBRN technical enablers during combined arms CWMD.

3-149. The following vignette provides an example situation for the integration of CBRN forces in support of stability. In this vignette, the situation (threat) is that a nuclear reactor in country X has been compromised and is emitting radiation. The incident exceeds the capacity of the local authorities, and country X has requested assistance to contain the incident. The United States has committed military personnel to support ICBRN-R.

Vignette Stability (ICBRN-R)

Mission: The CBRN task force has been given the mission to deploy to country X and provide ICBRN-R support. (See figure 3-15.) The task force, by order of the joint task force, will provide support to the U.S. Department of Energy and International Atomic Energy Agency personnel.

Organization: The proposed organization of a CBRNE task force includes one explosive ordnance disposal (EOD) company, four CBRN companies, two nuclear disablement teams, and one laboratory. Possible enablers could include medical, company, and interpreter support (not all inclusive). The command and support relationship is the CBRNE task force under the operational control of the geographic COCOM and is further assigned, as required.

Execution: In country, the CBRNE task force works for the joint task force, which in turn is in direct coordination with the United States Department of State. The CBRNE task force provides the integrating staff to plan, coordinate, and execute the interrelated CBRNE operational tasks. These tasks include conducting decontamination, providing CBRNE expertise, and responding to CBRNE incidents. The task force is predominantly composed of CBRN elements; however, in the event that the reactor compromise was a result of sabotage or terrorist activity, EOD elements may be requested by the host country and added to the task force. EOD elements would be prepared to respond to explosive hazards and to possibly training host-nation forces on EOD operations.

End state: Initial surveying and sample management are completed, with data turned over to the host nation. Decontamination support for responders and affected civilians is provided. Support is provided for possible improvised nuclear device and weapons incidents. Long-term surveillance and medical documentation for chronic health effects for Soldiers conducting CBRN operations is recommended.

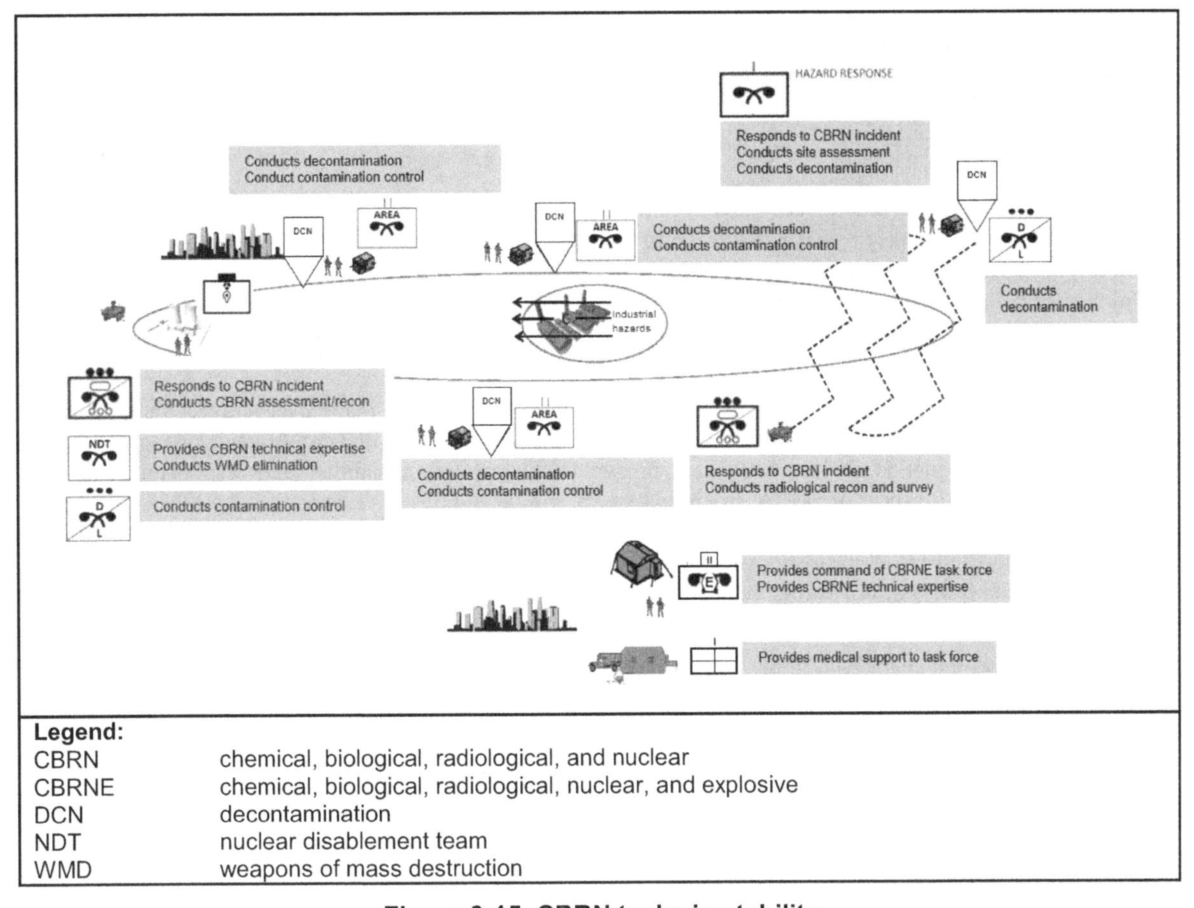

Figure 3-15. CBRN tasks in stability

3-150. Transitions mark a change of focus between phases or they mark the shift in relative priority between the elements of decisive action, such as from an offense to stability. Stability tasks include transitions of authority and control among military forces, civilian agencies, organizations, and the host nation. As stability is achieved, responsibility for response may be transferred to other military forces, governmental agencies, nongovernmental organizations, or local authorities. The civil affairs operations staff officer should be included in the planning and coordination of this transfer of responsibility. Determining whether DOD has met transition criteria requires close coordination and day-to-day interaction with the lead agency and includes G-9/battalion or brigade civil affairs operations staff officer (S-9) staff. This transfer must be carefully planned, coordinated, and executed with the relieving force or agency.

3-151. A transition to the consolidation of gains may occur, even if large-scale combat operations are occurring in other parts of an AO to exploit tactical success. A transition to the consolidation of gains may occur at the termination of large-scale combat operations and may encompass a lengthy period of postconflict operations before redeployment.

SECTION IV—CBRN CONTRIBUTION TO DEFENSE SUPPORT OF CIVIL AUTHORITIES

OVERVIEW

3-152. DSCA (as an element of decisive action) includes providing support for domestic disasters, domestic CBRN incidents, domestic civilian law enforcement agencies, and other designated support. Incidents involving CBRN material produce a chaotic and hazardous environment, requiring immediate response to minimize pain and suffering, reduce casualties, and restore essential infrastructure. Responders at the local, state, and federal levels may be overwhelmed by the magnitude of the incident, and U.S. DOD forces may be requested to provide additional support.

3-153. CBRN capabilities support the consequences of natural or man-made disasters, accidents, terrorist attacks, and incidents in the United States and its territories. Formerly called consequence management, the term CBRN response is characterized as a unique DOD response capability and responsibility. CBRN response is just one aspect of DSCA provided by U.S. military forces in response to requests for assistance from civil authorities for assistance. Domestic CBRN response is a type of support provided within the DSCA mission conducted by DOD forces to save lives; prevent injury; provide temporary critical life support; protect critical property, infrastructure, and the environment; restore essential operations; contain the event; and preserve national security.

3-154. Many other doctrinal references provide a source for additional information on DSCA, including—

- ADP 3-28.
- ATP 3-11.41.
- ATP 3-28.1.
- JP 3-41.

THREAT OVERVIEW

3-155. Threats may emanate from nation states or nonstate actors, such as transnational terrorists, insurgents, criminal organizations, and self-radicalized individuals. Within the spectrum of potential CBRN-related crises, a terrorist attack using WMD on U.S. soil presents daunting challenges for civilian authorities. Some unintentional CBRN releases, whether the result of accidents or natural causes, could create similar catastrophic results for civil authorities; however, attacks from people with sophisticated weapons knowledge can be much worse.

3-156. Terrorists have declared their intention to acquire and use CBRN agents as weapons to inflict catastrophic attacks against the United States. Extremist groups have a wide variety of potential agents and the delivery means from which to choose CBRN attacks. A terrorist's end goal is to use CBRN to cause mass casualties, panic, and disruption. Most likely, attacks will be on a small scale, incorporating a relatively crude means of delivery and easily produced or obtained chemicals, toxins, or radiological substances. The success of any attack and the number of ensuing casualties depend on many factors, including the technical expertise of those involved.

3-157. Terrorists have considered a wide range of toxic chemicals for attacks. Typical plots focus on poisoning foods or spreading agents on surfaces to poison via skin contact; however, some plots include broader dissemination techniques.

3-158. There are a wide range of TICs that, while not as toxic as blood, mustard, or nerve agents, can be used in much larger quantities to compensate for their lower toxicity. Chlorine and phosgene are industrial chemicals that are transported in multiton shipments by road and rail. Rupturing the container can easily disseminate these gases. The effects of chlorine and phosgene are similar to those of mustard agents.

Hazard Awareness and Understanding

3-159. Domestic CBRN response operations present unique challenges, working under differing legal authorities and chains of command when coordinating with and working alongside non-Department of Defense (DOD), state, local, and tribal agencies. The authorities for DOD components to conduct DSCA operations are found in DODD 3025.18 and other DODDs, standing CJCSs, U.S. Northern Command (USNORTHCOM) and U.S. Pacific Command (USPACOM) DSCA CONPLANS, and Headquarters, Department of the Army CBRN Response EXORD. Domestic CBRN response is managed at the lowest possible level, with DOD providing support, as directed.

3-160. To gain hazard awareness and understanding, the Army elements liaise and coordinate operations with interagency partners. Information gathered shapes the development of threat, hazard, vulnerability, and risk assessments and supports a better understanding of the complex OE.

3-161. Capabilities are aligned to support DSCA and specific CBRN response missions. The CBRN Response Enterprise consists of state National Guard and federal military forces. The CBRN Response Enterprise provides the nation with a dedicated, trained, ready, scalable, and tailorable military CBRN response capability.

Assess

3-162. Incident awareness and assessment addresses the limited information collection activities permitted in the homeland through consolidating information and providing analysis of the physical environment, weather impacts, terrorist threats, CBRN hazards, and other operational or mission variables.

3-163. During CBRN incidents in the homeland, CBRN assets identify, assess, advise, and assist higher commands and civil authorities.

- **Identify.** WMD–civil support team and CBRN reconnaissance elements detect, characterize, identify, and monitor unknown hazards by leveraging multiple detection technologies. Technical reachback to state and federal experts is used to support the identification process. The sophisticated detection, analytical, and protective equipment allows for operations in environments that may contain numerous, different CBRN hazards.
- **Assess.** Assessments occur with local, tribal, state, and federal response organizations before, during, and after an incident to ensure that CBRN elements are properly integrated into local and state emergency plans. They collect information from appropriate sources, identify pertinent data, and evaluate information to determine the mission threat, including hazards, risks, possible adversary actions, potential targets, the probability of an incident, the severity or level of the threat, and vulnerability to critical infrastructure.
- **Advise.** Supporting CBRN elements advises the incident commander and emergency responders on hazards and countermeasures. Advice may cover methods used during all phases of the operations to protect and mitigate the potential loss of life, damage to critical infrastructure, or extensive damage to private property. This advice assists emergency management authorities in tailoring their actions to minimize the impact of the incident.
- **Assist.** The CBRN elements assist the incident commander by providing expertise in hazard prediction modeling, liaison, downrange survey operations, hazard mitigation, and recovery planning.

Protect

3-164. Protection capabilities for military forces against CBRN hazards apply domestically, but in some circumstances they must adhere to certain U.S. codes. During a CBRN incident, the civilian incident command staff normally determines the level of protective garment required based on the hazard.

3-165. Units should give special attention to heat injury prevention for Soldiers using personal protective equipment. Additionally, they should be aware of toxic industrial materials in the affected areas.

3-166. USNORTHCOM and USINDO-PACOM and subordinate commanders, when tasked by the Secretary of Defense, may provide protection capabilities during homeland defense and domestic CBRN response operations. National Guard forces have similar protections when functioning on state or federal active duty.

MITIGATE

3-167. State National Guard forces and federal military forces are prepared to respond to CBRN incidents as part of an overall DOD CBRN response enterprise. CBRN incidents may involve a response to a single incident site or to multiple sites in different states. DOD may commit some or all of a standing joint task force and a defense CBRN response force. This consists of a tiered response of CBRN assets that includes weapons of mass destruction–civil support teams (WMD-CST), CBRN Enhanced Response Force Package (CERFP), Homeland Response Forces (HRF), Defense CBRN Response Force (DCRF), and Command and Control Response Elements (C2CRE). Other specialized CBRN units could include technical support forces and Defense Threat Reduction Agency teams. In CBRN incidents, DOD installations serve as staging areas for resources and agencies.

3-168. It is imperative for a rapid and effective employment of reconnaissance capabilities to provide assessments on the effects in terms of casualties and medical treatment (detect and monitor). These assessments provide the necessary information to assist the incident commander in determining upwind and crosswind points and best locations for search and rescue, decontamination, medical triage, emergency medical services, and other sites. DOD forces reinforce evacuation centers to increase capacity and throughput or provide forces to conduct search and rescue, casualty decontamination, medical triage, and emergency medical stabilization.

DEFENSE SUPPORT OF CIVIL AUTHORITIES PLANNING CONSIDERATIONS

3-169. Important elements of planning considerations—when in support of civil authorities—include a good understanding of the OE; authorities; policies; laws; hazard and threat awareness; critical infrastructure; available capabilities that support a CBRN response; and medical, logistical, and intelligence resources.

3-170. The development of the OE includes considerations of the geographical terrain, climate, population, infrastructure, and jurisdictional authorities. Most state, local, tribal, and territorial agencies plan to use local and state resources, Emergency Management Assistance Compact (EMAC) capabilities, and the Initial Response Authority of local T10 units before incorporating Title 10 resources into the plan.

3-171. It is important for Army planners to understand that, to be able to conduct life-saving operations during a CBRN incident, they must be on the ground and operational within 24 hours to maximize their effectiveness and contribute to saving lives, which occurs within the initial 72 hours after the incident.

3-172. Integrated planning between the Army Service component command, Headquarters, Department of the Army, United States Army Forces Command, and allocated units ensures the development of operational plans that nest effectively with COCOM CONPLAN requirements. Even though locations are different, much of the detailed planning information is transferable to other scenarios in other locations, perhaps with only minor refinements.

3-173. A catastrophic CBRN incident resulting in numerous casualties and the disruption of normal life-support systems will overwhelm the capabilities of local, state, and other federal agencies. Once that happens, the primary agency will submit a request for assistance (RFA) through FEMA to DOD within hours following a catastrophic CBRN incident.

Appendix A
Command and Support Relationships

A-1. Because of the complexity of OE and METT-TC requirements, CBRN elements are task-organized to meet the requirements of an assigned mission. Task organization is routinely multi-component and can include any combination of platoons, companies, and battalions organized under the task force headquarters.

A-2. Commanders use command and support relationships to build combined arms organizations. Command relationships define command responsibility and authority. Support relationships define the purpose, scope, and effect desired when one capability supports another. Command and support relationships are fundamental to effective CBRN support to operations. Operations orders that place units under the command of a different headquarters for any length of time must include a detailed summary of the relationship between the unit, its new headquarters, and its parent unit. Typically, the smallest detached CBRN force is a platoon.

A-3. The CBRN commander requires freedom to employ forces to adapt to the current OE, but traditionally the command relationship for CBRN elements brigade and above is attached or in direct support. Figure A-1, page A-2, depicts traditional relationships for brigade and above elements. The joint task force is typically in direct support to the theater or Army, and the CBRN brigade is most likely in direct support to the corps or possibly in general support to allow for the freedom to adapt to mission requirements. The maneuver commander requests needed capabilities, and the CBRN staff assists with determining the right task organization of CBRN assets to support.

A-4. CBRN elements battalion and below are organized to provide the best support to the maneuver commander based on the tactical situation. A CBRN battalion in support of the division may be further task-organized with tactical control of companies and platoons to maneuver elements. The CBRN battalion commander can advise the maneuver commander on the best employment of the subordinate CBRN elements. The best employment practice maintains integrity of platoons and specialty teams to maximize their capability. Figure A-1, page A-2, depicts the preferred relationship of CBRN elements battalion and below. See chapter 3 for more information on the employment of CBRN elements in support of decisive action.

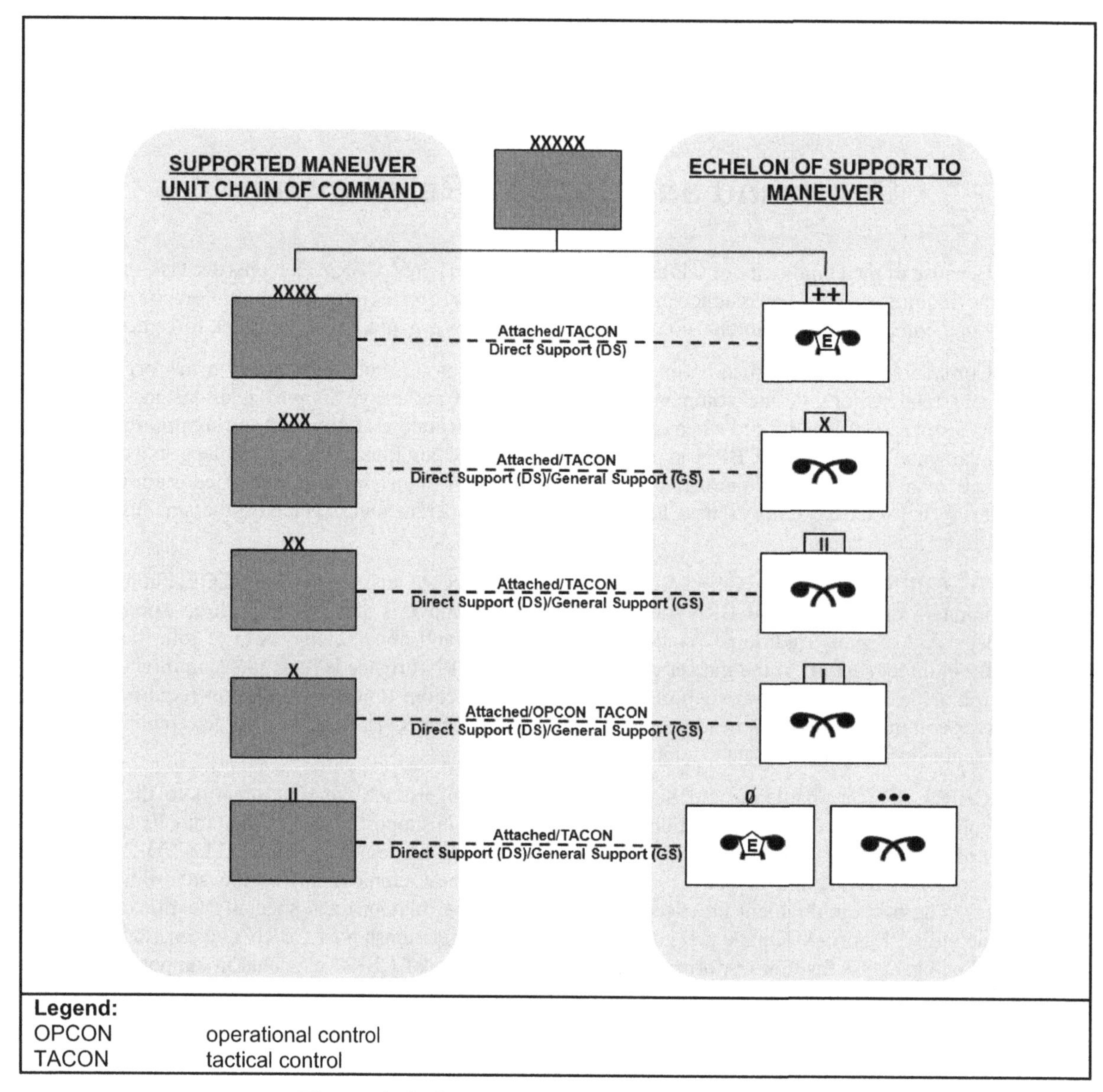

Legend:

OPCON	operational control
TACON	tactical control

Figure A-1. CBRN support brigade and above

Appendix B
Warfighting Function Considerations

CBRN SUPPORT TO WARFIGHTING FUNCTIONS

B-1. Warfighting functions are used to generate combat power and provide the commander a way to conceptualize capabilities. There are eight elements of combat power—leadership, information, mission command, movement and maneuver, intelligence, fires, sustainment, and protection. The commander uses leadership and information to integrate capabilities to generate combat power. The six warfighting functions are not specific to an organization or branch but are an intellectual way of understanding how groups of tasks and systems come together to provide an element of combat power. There are unique considerations within each of the CBRN warfighting functions.

MISSION COMMAND

B-2. The *mission command warfighting function* is the related tasks and systems that develop and integrate those activities enabling a commander to balance the art of command and the science of control in order to integrate the other warfighting functions (ADP 3-0). The mission command warfighting function is unique in that it integrates the activities of the other warfighting functions.

B-3. CBRN units must integrate mission command and the operations process activities for the unit while interacting with the mission command activities of the unit being supported. The interaction may be primarily through a CBRN staff assigned to the supported unit or through staff counterparts. In some cases, a supported unit may not have assigned CBRN staff; therefore, the supporting unit will provide this support as well. This relationship and the degree of integration are determined by many factors, including the type of unit and echelon being supported and the command or support relationship established. For example, an ABCT may have a hazard response company TACON to it. If so, the brigade CBRN officer and hazard response company commander would collaboratively plan, coordinate, and synchronize CBRN efforts across the brigade. The hazard response company could be placed under the BEB for command authority. The BEB may control and direct the application of the hazard response company's capabilities while providing sustainment support.

B-4. CBRN units are generally task-organized as members of combined arms teams. Because the available CBRN units are designed for specific tasks, capabilities must be shifted within the AO to match the requirements with the capabilities of modular CBRN units. Transitions occur at the strategic, operational, and tactical levels, and flexibility in the task organization is required to permit the shifting of CBRN capabilities.

B-5. Control measures can be tools to help units identify key points. Decontamination points, NAIs with potential for TIMs that may need to be designated as no-fire areas, and coordination points for linkup between units and CBRN elements are examples for consideration. Postincident, CBRN staffs mark contaminated areas and distribute information to all for situational awareness.

B-6. CBRN staff work with medical, public affairs, and information operations personnel to achieve the commander's communication and other nonlethal objectives. One measure of success is how well CBRN staffs work together with these staff sections to develop and disseminate an uninterrupted flow of information on the implications of WMDs in the AO. This information flow, often disseminated through Solider and leader engagements, can be used to shape and influence foreign populations by expressing information subjectively to influence perceptions and behaviors and to obtain compliance, noninterference, or other desired behavioral changes.

MOVEMENT AND MANEUVER

B-7. The *movement and maneuver warfighting function* is the related tasks and systems that move and employ forces to achieve a position of relative advantage over the enemy and other threats (ADP 3-0). CBRN units and staffs support movement and maneuver warfighting functions through the Chemical Corps core functions.

B-8. CBRN supports the movement and maneuver warfighting function through mobility and survivability. CBRN reconnaissance supports mobility operations by locating and marking contaminated areas and routes, allowing maneuver forces to avoid unnecessary exposure. Providing protection from CBRN hazards allows maneuver forces to fulfill their primary mission. Decontamination increases the ability of maneuver forces to withstand CBRN conditions in the environment.

B-9. Operating in close combat support to maneuver forces requires that CBRN reconnaissance elements have the ability to integrate and coordinate actions with fire, movement, or other actions of combat forces. CBRN reconnaissance should not be held in reserve and should be weighted in efforts to identify potential threats before they become a hazard and give the command decision space. CBRN reconnaissance capabilities are in limited supply; therefore, when considering their employment, the commander must also weigh the costs and benefits of employing this limited asset in terms of the battlefield framework.

B-10. CBRN forces add to the combat power of movement and maneuver in assessing clean and dirty routes through the battlespace. The employment of CBRN reconnaissance elements reduces the risk of traveling through contaminated areas and spreading contamination. CBRN forces give the commander options for understanding required protection for the hazard and to reduce the impact protective equipment has on the speed of movement.

INTELLIGENCE

B-11. The *intelligence warfighting function* is the related tasks and systems that facilitate understanding the enemy, terrain, weather, civil considerations, and other significant aspects of the OE (ADP 3-0). CBRN forces make a critical contribution to this warfighting function through the core function of assessing CBRN threats and hazards and the integrating activity of hazard awareness and understanding. CBRN staff and planners provide a predictive and deductive analysis of enemy CBRN capabilities to intelligence.

B-12. CBRN staffs contribute the knowledge and understanding of CBRN threats and hazards. They assist the G-2/S-2 in developing an understanding of enemy CBRN capabilities. They work with G-2/S-2 during planning to analyze potential threats and to evaluate how the enemy might use CBRN hazards to impact operations. They collaborate with the G-2/S-2 to provide estimates for the unconventional use of TIM to create hazards for U.S. forces. They advise the commander on the influences that terrain and weather have on CBRN hazards.

B-13. CBRN reconnaissance provides data and information that contributes to answering PIR. CBRN reconnaissance elements contribute information about CBRN hazards through their information collection efforts. Specialized CBRN assets may need to be available to collect the information needed to answer these requirements. Reconnaissance teams are focused on the collection of tactical and technical information to support the BCT freedom of maneuver and the survivability of friendly forces and facilities.

B-14. CBRN staffs at the division, corps, and theater army echelon and in-theater CBRN headquarters determine CBRN-related intelligence requirements in a potential AO. They collect and analyze CBRN-related intelligence data in coordination with the respective G-2.

B-15. Effective offensive and defensive actions capitalize on accurate, predictive, and timely intelligence. IPB is an integrating process of the operations process. IPB results in the creation of intelligence products that are used during MDMP to aid in developing friendly COAs and decision points for the commander. The G-2/S-2 leads the IPB process. The CBRN officer is the staff subject matter expert on CBRN and assists the G-2/S-2 in determining the locations of threat CBRN assets and potential areas of employment. Conclusions reached during IPB are critical to planning information collection and targeting operations.

B-16. The G-2/S-2 staff provides the most likely (and the most dangerous) COAs based on threat intent and capabilities. CBRN staffs are involved in the IPB process with intelligence sections, providing input into enemy CBRN capabilities, release authorities, terrain effects on CBRN use, enemy doctrine, and COAs. CBRN input into CWMD specific intelligence, information operations, civil military operations, and civil affairs operations. This input feeds PMESII-PT; areas, structures, capabilities, organizations, people, and events (ASCOPE) analysis; and sewage, water, electricity, academics, trash, medical, safety, other considerations (SWEAT-MSO), which provides a holistic approach to understanding a complex environment by analyzing the factors and systems that influence such environments. These intelligence products are then added to the existing targeting packets. By doing so, the staff establishes an effective process to integrate and fuse all sources of available threat information to provide for a continuous analysis of threat information that identifies the full range of known or estimated terrorist threat capabilities, intentions, and current activities.

B-17. The CBRN staff lends insight to this process from a CBRN perspective. The CBRN staff provides insight into the development of CCIR, which shapes information collection activities, including CBRN R&S tasks and purposes.

FIRES

B-18. The *fires warfighting function* is related tasks and systems that provide collective and coordinated use of Army indirect fires, air and missile defense, and joint fires through the targeting process (ADP 3-0). CBRN capabilities significantly contribute to this warfighting function when they are used to facilitate targeting.

B-19. The integration of CBRN and fires can provide valuable information to understanding the enemy's WMD capabilities. Information from radar analysis can aid in identifying missiles filled with chemical agents. CBRN staffs must provide advice to planning fires so as not to cause collateral damage that increases the hazard.

B-20. CBRN staffs provide the subject matter expert for the development of targeting packets during the targeting process. When the potential to encounter WMD sites exists, CBRN SMEs provide information on what potential materials may be found at a site, the threats posed by the materials at the site, and the impact of these threats on future operations. The CBRN staffs advise the targeting team on the impacts of WMD employment and on targeted storage or production sites. CBRN staffs must be prepared to provide data for COA development (assess COA feasibility, acceptability, and suitability). See figure B-1, page B-4, for an example of COA development.

B-21. CBRN staffs fuse developing intelligence with the knowledge of available CBRN capabilities. Integrating this information provides an analysis of which CBRN reconnaissance and response capabilities are available to collect information and respond to enemy actions. During the targeting cycle, CBRN staff provides data and makes recommendations for further actions. They continually refine and modify COAs with additional information collected from the running staff estimate. The updated running staff estimate prepares the way to start the next targeting cycle.

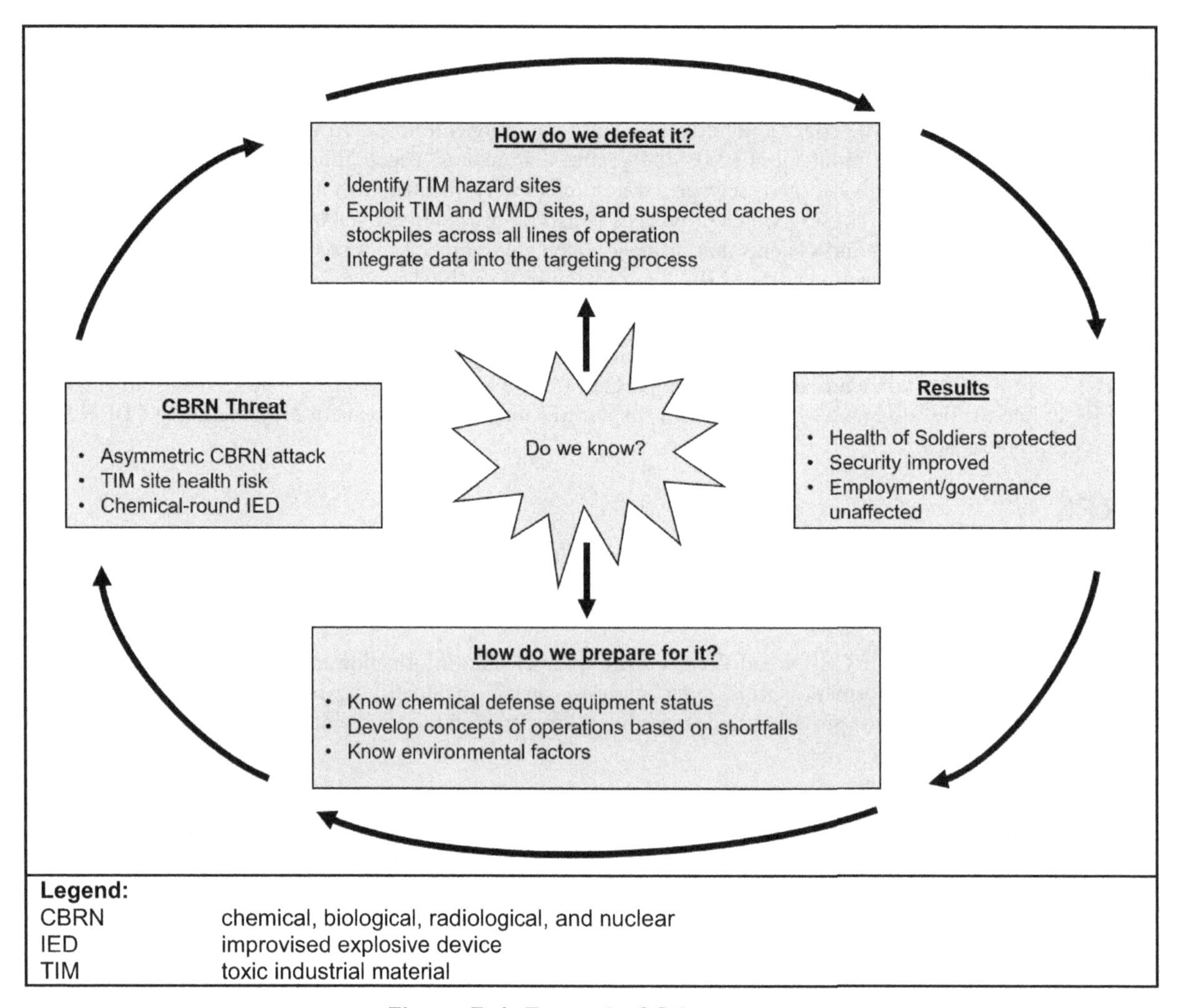

Figure B-1. Example COA process

B-22. The targeting cycle is an example of an enduring MDMP with an existing campaign plan, lines of operation, and the commander's guidance. Each staff section or warfighting function develops a running staff estimate, which is a tool used to conduct the initial mission analysis (MA). For effective CBRN MA, CBRN staffs must develop a running staff estimate for the OE that includes current CBRN threats, such as weaponized CBRN material, TIMs, nonweaponized biological material, estimated templated CBRN strikes, and improvised explosive devices and vehicle-borne improvised explosive devices with chemical accelerants or conventional attacks against a TIM facility. Terrain and weather conditions in the OE are identified, as required. The completion of the running staff estimate leads to the mission analysis brief to the commander. See appendix D, figure D-6, page D-5, for an example of a CBRN running staff estimate.

Note. See ATP 3-60 for more information on the targeting process.

SUSTAINMENT

B-23. The *sustainment warfighting function* is the related tasks and systems that provide support and services to ensure freedom of action, extend operational reach, and prolong endurance (ADP 3-0). Continuing operations in CBRN environments creates a reliance on sustainment capabilities. The sustainment of protection capabilities (MOPP suits and filters for COLPRO), consumables for CBRN detection, and identification equipment requires detailed planning. CBRN operations require intense water sustainment, resupply of decontaminants, vehicle replacement parts, and medical chemical defense material.

B-24. The lowest echelons must report readiness by using chemical defense equipment reports. These reports incorporate information obtained at in-processing, including IPE sizing, optical inserts, equipment readiness, manning, and training. The information is forwarded to the battalion level for consolidation and is submitted alongside the Defense Readiness Reporting System–Army. The Defense Readiness Reporting System–Army gets further consolidated at higher echelons.

B-25. Additionally, medical and sustainment units must be postured to support units conducting detailed troop decontamination, and mortuary affair units must be postured to process contaminated human remains through the mortuary affairs contaminated remains mitigation site. Planning between CBRN forces and mortuary affairs units occurs to ensure the duplication of fatality collection points so that contaminated and uncontaminated fatalities are not mixed. The logistics ability to resupply impacts the CBRN capability to protect or conduct decontamination to sustain in contaminated environments. Requirements for the repair and replacement of systems increase during large-scale combat operations.

Note. See JP 4-0 for additional information on mortuary affairs planning.

B-26. Sustainment operations are challenged in large-scale combat operations in CBRN environments due to the frequent cross contamination of MSRs and alternate supply routes and to the enemy targeting of large sustainment nodes. It is critical to the success of sustainment operations that sustainment units frequently incorporate CBRN conditions in METs.

B-27. The health service support mission integrates with CBRN to support the treatment of CBRN casualties. Contaminated casualties can put the entire health service support system at risk if the proper precautions are not taken to prevent the transfer of contamination. Planning between CBRN forces and health service support occurs to ensure the duplication of casualty collection points so that contaminated casualties do not mix with uncontaminated casualties.

PROTECTION

B-28. The *protection warfighting function* is the related tasks and systems that preserve the force so the commander can apply maximum combat power to accomplish the mission (ADP 3-0). Conducting CBRN operations is one of the key protection tasks. Tasks and systems of CBRN units and staffs are linked to the protection warfighting function. For example, the CBRN staff at a brigade must be able to execute the task (which is to prepare the brigade for operations under CBRN conditions) to ensure that the brigade combat power is not degraded during a CBRN attack.

B-29. Force health protection and CBRN must coordinate efforts to promote, improve, or conserve the health of Soldiers when the threat of CBRN hazards exists. The functions of force health protection and CBRN intersect in areas of preventative medicine, medical surveillance, veterinary services, and laboratory support. CBRN and medical personnel combine their knowledge to provide the commander advice on preventative medicine measures that can be taken to reduce risk to the medical effects of chemical, biological, or radiological hazards. The environmental and biological surveillance of outbreaks can provide indicators of attack. Laboratory support for environmental samples taken by CBRN reconnaissance teams requires prior planning.

PROTECTION IN OFFENSIVE OPERATIONS

B-30. During the four types of offensive operations (movement to contact, attack, exploitation, and pursuit), CBRN capabilities assess threats and hazards, protect against CBRN environments, and mitigate hazard effects. Commanders develop a scheme of protection for the transition of each phase of an operation or major activity. Transitions mark a change of focus between phases or between the ongoing operation and execution of a branch or sequel. Shifting protection priorities between offensive, defensive, and stability tasks also involves a transition. Transitions require planning and preparation well before their execution so that a force can maintain the momentum and tempo of operations.

B-31. During the offense, CBRN battalions are assigned within the division support area to provide sustainment of subordinate units and oversight of laboratory support, as required. Hazard response companies provide CBRN route reconnaissance from the brigade support area to the forward edge of the battle area. They also provide thorough decontamination support, as required. Maneuver units must use organic decontamination assets for operational decontamination, but they can request CBRN units to augment personnel to support them. Brigade organic CBRN reconnaissance units conduct screening within the enemy disruption zone to provide early warning of CBRN at NAIs.

B-32. Accomplishing movement and maneuver in a CBRN environment is difficult and, in some situations, the commander may direct movement and maneuver to avoid areas contaminated by CBRN elements. Preserving combat power from the effects of CBRN incidents is essential for the commander to seize, retain, and exploit the initiative.

Protection in Defensive Operations

B-33. All units have an inherent responsibility to improve the survivability of their own fighting positions, bases, or base camps. CBRN personnel contribute to unit protection by performing vulnerability assessments. These assessments provide a list of recommended activities actions ranging from CBRN protection to contamination mitigation for commanders to consider—

- **Vulnerability assessments.** CBRN vulnerability assessments provide insight into the ability of the unit to mitigate likely CBRN events and prompt the unit to develop procedures, acquire equipment, and take actions to correct vulnerabilities.
- **COLPRO.** In preparing for the defense, CBRN staffs must consider plans for the use of COLPRO. They must consider the—
  - Limitations.
  - Resources required.
  - Necessary preparations.
- **Warning and reporting.** Early warning of CBRN hazards alerts that an attack has occurred so that the correct protective measures can be taken.
- **Decontamination.** Preparation for operations in a CBRN environment requires planning for immediate or operational decontamination. Battle drills for the immediate measures to take after CBRN attacks occur increase the force ability to survive CBRN conditions.

Threat Reduction

B-34. Threat reduction cooperation includes those activities undertaken with the consent and cooperation of host-nation authorities in a permissive environment to enhance physical security and to reduce, dismantle, redirect, and/or improve the protection of a state's existing WMD program, stockpiles, and capabilities. Tactical commanders provide threat reduction cooperation activities in support of CWMD objectives. The principle purpose of these activities is to deny rogue states and terrorists access to weapons, material, and expertise. Other states may need assistance with more discrete requirements to dismantle or destroy WMD in excess of defense needs; to comply with international treaty obligations (such as the Chemical Weapons Convention); or to impose export control, border control, law enforcement, and antismuggling capabilities.

B-35. Threat reduction cooperation also responds to opportunities to roll back or eliminate a state's WMD programs and capabilities on cooperative terms; for example, Libya's decision to voluntarily dismantle its WMD programs. Another challenge is the safety and security of WMD inventories of friendly or nonhostile states. Existing security arrangements may be viewed as inadequate to prevent theft, sabotage, or accidental release. Threat reduction cooperation occurs in a permissive environment, and while they are not primarily a COCOM responsibility, combatant commands must maintain visibility of these efforts to ensure that theater security cooperation plans and security measures are consistent with threat reduction initiatives. The following military tasks directly or indirectly support threat reduction cooperation in a permissive environment:

- Provide security for current WMD, related materials, and systems from theft, sabotage, or unauthorized use.
- Support efforts to ensure the safety of WMD and delivery systems from accidental or inadvertent release.
- Maintain situational awareness of WMD safety and security issues, and communicate concerns to senior leaders.
- Integrate the commander's safety/security concerns and threat prioritization with operational level guidance.
- Assign responsibilities for threat reduction cooperation, and coordinate efforts with other commands.

B-36. Tactical nonproliferation activities are not conducted sequentially and discretely in the prosecution of tactical level military operations, but they will occur independently or simultaneously in response to security cooperation and partner activities and threat reduction cooperation. Tactical commanders should be prepared to provide short-notice support to cooperative WMD threat reduction efforts. Supporting tasks that are directly or indirectly related to cooperative WMD threat reduction efforts include emplacing sensors and conducting monitoring, detection, and security operations.

This page intentionally left blank.

Appendix C

Domestic and International CBRN Response

CBRN RESPONSE

C-1. Regardless of the lead agency for a CBRN response effort, CBRN units and staffs provide support through the core functions of assessing threats and hazards, providing protection, and mitigating CBRN incidents. Figure C-1 depicts the response environments.

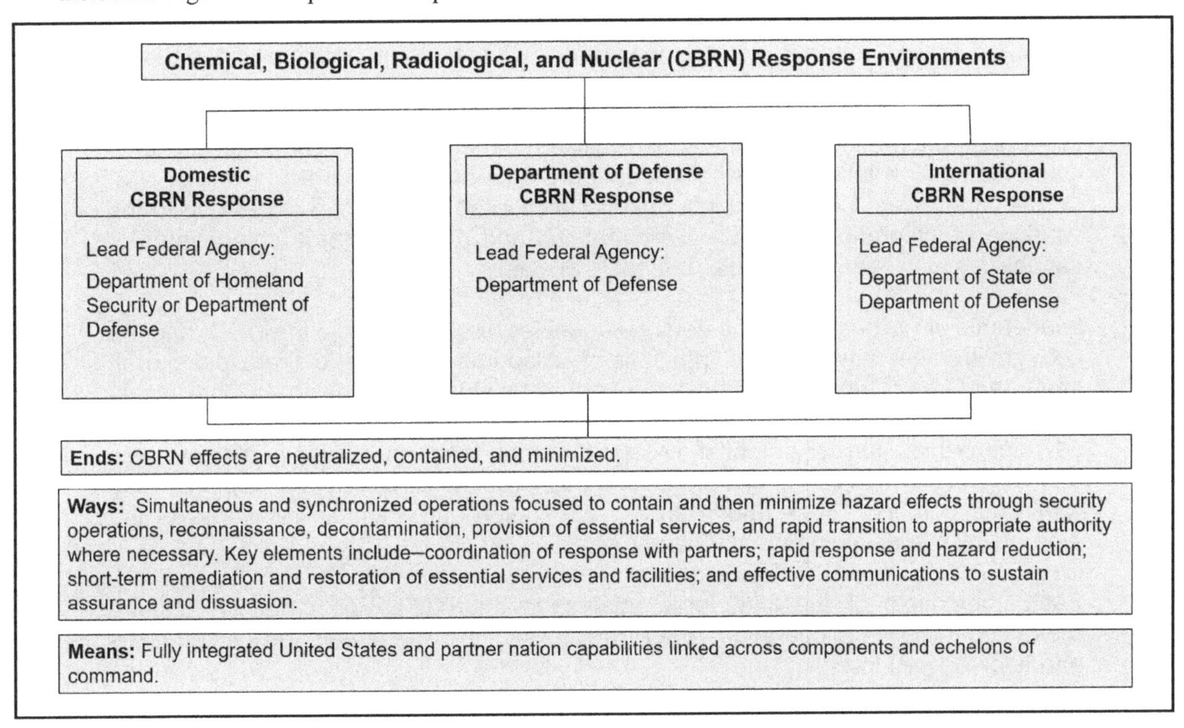

Figure C-1. CBRN response environments

DOMESTIC RESPONSE

C-2. CBRN capabilities support the consequences of natural or man-made disasters, accidents, terrorist attacks, and incidents in the United States and its territories. Formerly called consequence management, the term CBRN response is characterized as a unique DOD response capability and responsibility. CBRN response is just one aspect of DSCA provided by U.S. military forces in response to requests for assistance from civil authorities for assistance. Domestic CBRN response is a type of support provided within the DSCA mission conducted by DOD forces to save lives, protect property and the environment, and meet basic human needs.

INTERNATIONAL RESPONSE

C-3. ICBRN-R is assistance provided by the U.S. government to an impacted nation to respond to the effects of a deliberate or inadvertent CBRN incident on foreign territory. DOD CBRN response includes immediate life-saving measures for the affected host-nation population, U.S. citizens, armed forces abroad, and its friends and allies to minimize human casualties and provide temporary associated essential services.

ICBRN-R applies to international incidents involving the deliberate or inadvertent release of CBRN materials, including TIMs.

Vignette

Domestic and international CBRN response discussions have moved to center stage nationally and internationally, as reflected in the National Security Strategy 2017. CBRN events have transitioned from a footnote in history to a plausible and viable domestic and international threat. This transition has triggered—in leadership's mind—the necessity and significance of interoperability within allies, interagency, and the joint force, with the intent of solidifying methodology, collaboration, and agreements to address CBRN events.

Allies and coalition partners in the United States Central Command and United States Pacific Command AORs have precipitated a discussion with leaders in the U.S. military and Department of State due to threat, demonstration, and/or use of CBRN material in the region. International diplomatic and military leadership intent on protecting their countries foster relationships with the United States and/or acquire a myriad of U.S. CBRN capabilities. Considerations must be accounted for with U.S. allies due to the variance of integration of U.S. military material and doctrine. Some components or capabilities may be interoperable, but not in all cases.

Leadership of various homeland defense agencies, regions, states, the DOD, and the U.S. government have pursued variations of collaborative forums to share ideas, draft memorandums of understanding, and conduct training exercises to validate plans. Components of the interagency acknowledge NSS directives; the significance of advance preparation for potential threats; the importance to efficiently mitigate pain and suffering of the American people during an event; and reduction of vulnerabilities within our borders. These efforts protect the Nation and maintain readiness of the emergency response support infrastructure.

Each component of the joint force retains an internal CBRN capability. During protracted unified land operations, the Army provides the way ahead and synchronizes efforts of the joint force.

Appendix D
CBRN Staff

CBRN CONTROL CENTER

D-1. CBRN cell functions are to analyze, plot, and advise the commander on actions in CBRN environments. Capabilities for CBRN control exist all the way down to the company level. The first echelon resourced to conduct continuous operations is the division. This cell is required to report CBRN messages in all formats (voice and digital). The CBRN warning and reporting system allows commanders and CBRN staffs to determine required contamination avoidance measures and to plan operations accordingly. The reports are scaled (CBRN 1-6) in terms of information and time.

D-2. The CBRN 1 observer's initial report stops at the CBRN control center or is sent to higher echelons as requested. The CBRN 1 report is created by the initial observer of the CBRN incident and is sent through mission command nodes from the controlling unit headquarters to the division control center. It may be sent to higher echelons if requested. Figure D-1 depicts the routing of a CBRN 1 report.

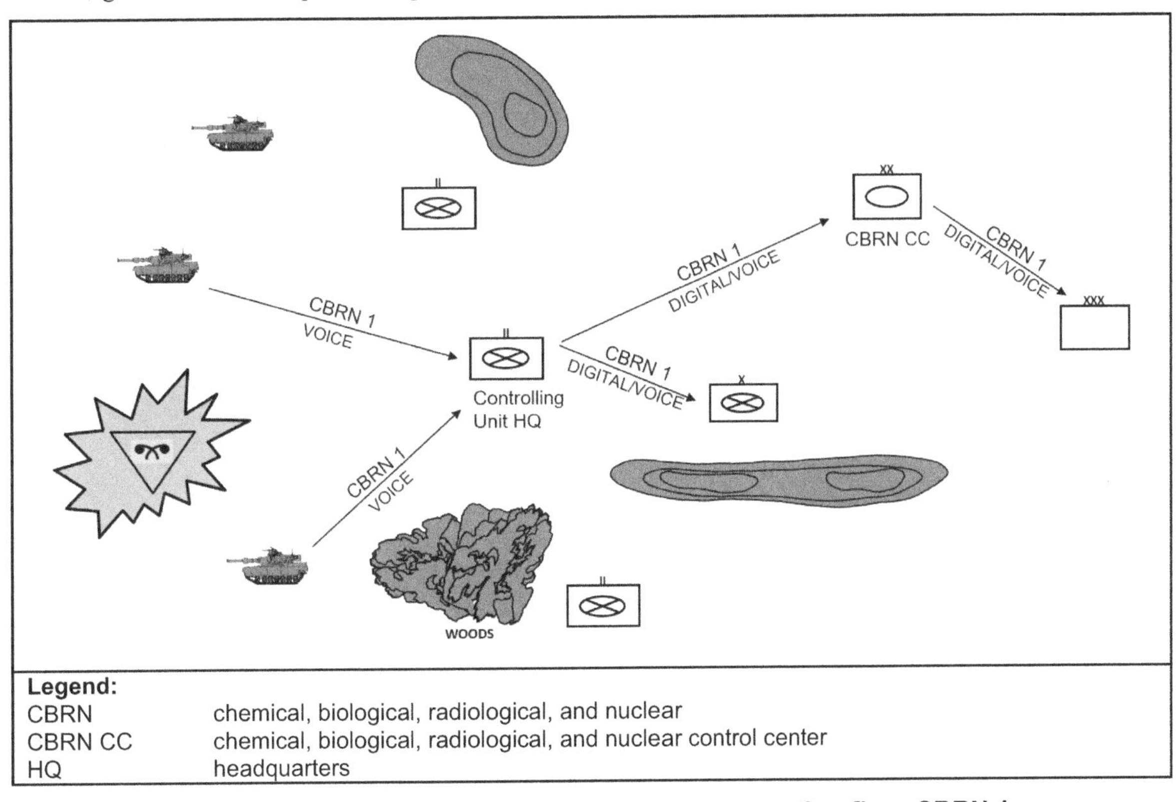

Figure D-1. CBRN event immediate response reporting flow–CBRN 1

D-3. The CBRN 2 evaluated data report is completed by the controlling headquarters to be distributed laterally to adjacent units and to be submitted to the division control center to prevent cross contamination during unit maneuver. Figure D-2 depicts the flow of the CBRN 2 report.

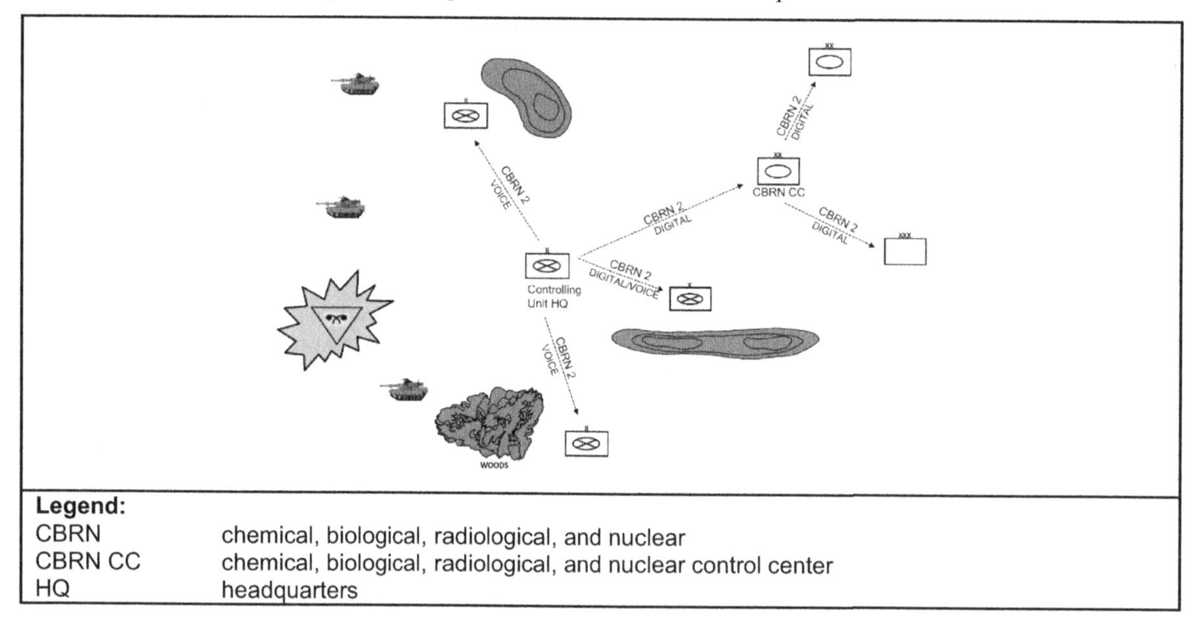

Figure D-2. CBRN event evaluated data reporting flow–CBRN 2

D-4. The CBRN 3 immediate warning of expected contamination or hazard area report is disseminated to all units that may be affected. It is primarily prepared and distributed by the division CBRN control center. The battalion and brigade have the capability to prepare a CBRN 3 report, but they may not have the time or capacity to prepare and distribute it. A subordinate unit may just receive the order to move instead of a full hazard area plot. See figure D-3 for an example of the flow of a CBRN 3 report.

Note. More details of reporting at higher echelons can be found in STANAG 2103.

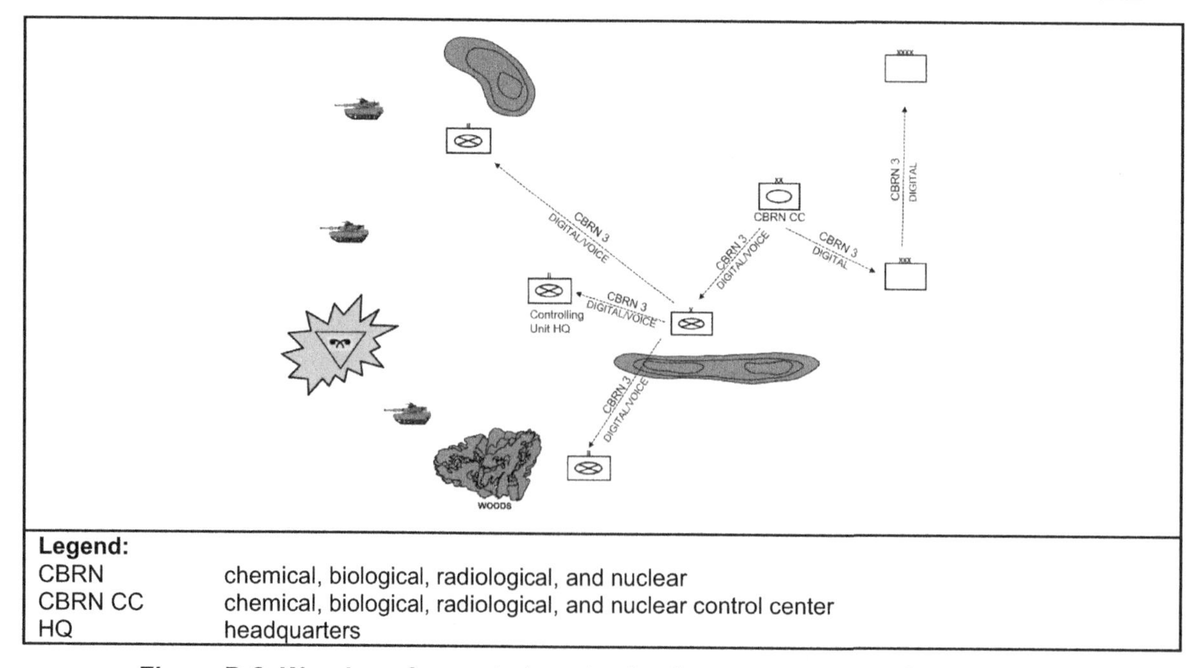

Figure D-3. Warning of expected contamination area reporting flow–CBRN 3

D-5. The CBRN 4 reconnaissance, monitoring, and survey report is sent to the division CBRN control center to report specific contaminated areas from reconnaissance, survey, and monitoring data. The data from reconnaissance and surveillance must be reported to the division CBRN control center for processing, but the communications may route through the chain of command for situational understanding. The CBRN 4 report is also used by units to report contaminated areas left by decontamination. Figure D-4 depicts the flow of the CBRN 4 reports.

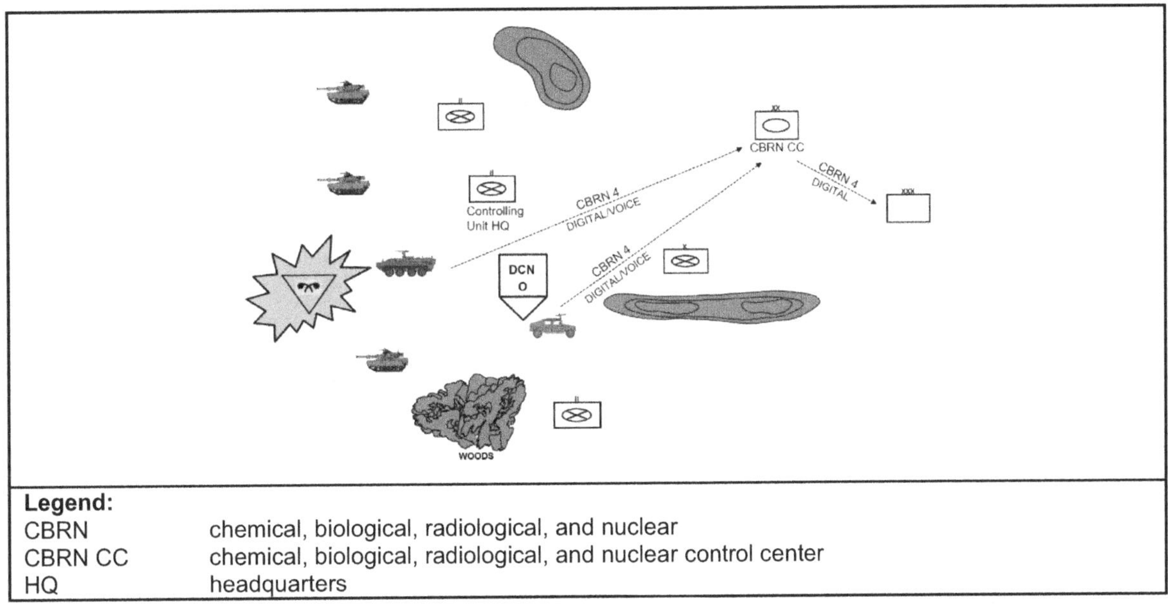

Legend:
CBRN chemical, biological, radiological, and nuclear
CBRN CC chemical, biological, radiological, and nuclear control center
HQ headquarters

Figure D-4. Reconnaissance, monitoring, and survey results reporting flow–CBRN 4

D-6. The CBRN 5 areas of actual contamination report is used for areas of actual contamination and is prepared by the CBRN control center to be distributed to major subordinate commands and to corps and theater operations centers. A CBRN 6 detailed information of CBRN report is prepared as needed and is not depicted. Figure D-5 depicts the flow of the CBRN 5 report.

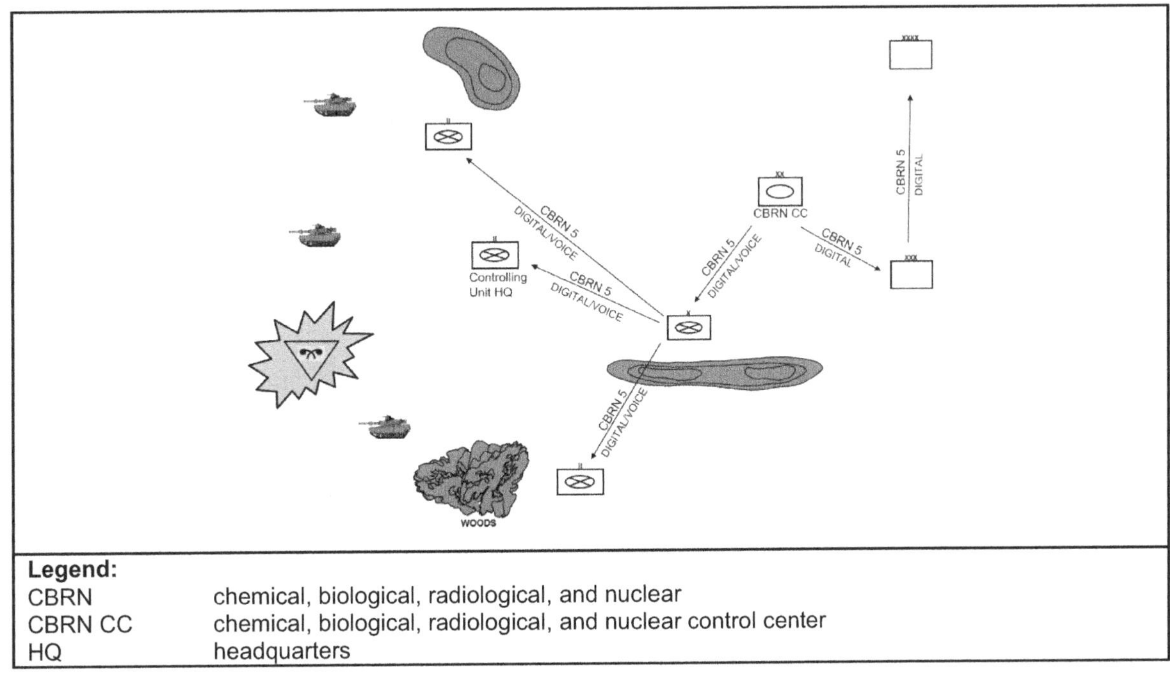

Legend:
CBRN chemical, biological, radiological, and nuclear
CBRN CC chemical, biological, radiological, and nuclear control center
HQ headquarters

Figure D-5. Areas of actual contamination reporting flow–CBRN 5

PROTECTION AND PROTECTION WORKING GROUP

D-7. The protection working group plans, coordinates, integrates, and synchronizes protection tasks and systems for each phase of an operation. At division and higher, the CBRN staff integrates CWMD operations, defense, CBRN response, WMD tactical disablement, and disposal. At brigade level and below, the integration occurs more informally, with the designation of a protection coordinator from among the brigade staff or as an integrating staff function assigned to a senior leader.

PLANNING

D-8. Initial assessment supports protection prioritization, threats and hazards, criticality, and vulnerability. The scheme of protection describes how protection tasks support the commander's intent and concept of operations, and it uses the commander's guidance to establish the priorities of support to units for each phase of the operation.

D-9. The protection working group uses information derived from commander's guidance, the IPB, targeting, risk management, warning orders, the critical and defended asset list (division and higher), and the mission analysis to identify critical assets.

D-10. Corps and division protection working groups coordinate closely with staff personnel to identify information and assets that need protection and apply appropriate protection and security measures consistent with their collective threat analysis.

D-11. Commanders at all echelons maintain the effectiveness of their force in CBRN environments by establishing CBRN defense plans that—

- Estimate enemy intent, capabilities, and effects for CBRN.
- Provide guidance to the force on necessary protective measures.
- Apply the IPB output to develop CBRN reconnaissance plans to answer PIRs.
- Establish the employment criteria of CBRN enablers to counter CBRN threats.
- Establish a logistic support plan for long-term CBRN operations.
- Establish CBRN warning and reporting requirements.

D-12. CBRN defense plans provide detail to subordinate units, providing preattack preparations and postattack execution measures. Staffs at every echelon must be familiar with the mission, capabilities, and current situation to ensure that their assessments and recommendations provide meaningful operations for action by the commander.

D-13. Conducting IPB is fundamental to understanding enemy CBRN capability and intent. The IPB process must account for confirmed and plausible enemy capabilities, plans, and actions. When focused on CBRN effects, the information collection plan will prioritize CBRN reconnaissance assets to areas of greatest importance.

D-14. Overall, success in a CBRN environment depends on the effective integration of CBRN equipment; training; and CBRN tactics, techniques, and procedures while preparing for and executing operations.

PREPARATION

D-15. The protection working group ensures that controls and risk reduction measures developed during planning have been implemented and are reflected in plans, standard operating procedures, and running estimates (see figure D-6), even as threat assessments are continuously updated.

Situation

Weather effects: How the current/projected weather will affect chemical agents and assets

Terrain effects: How the current/projected terrain will affect chemical agents and assets

Enemy capabilities:	Chemical agents:	Delivery systems:

Templated strikes: (location, type, time)

Known strikes: (location, type, time)

TIM facilities	Industry	TIM	Location	Hazard
Brief updated information on security and site assessments				

MOPP

Decon
Task
Purpose
Commander's intent

Linkup points
Vehicle decon
Chemical casualty collection point

Map of AO with chemical-related graphics

Chemical assets

Recon	**HRC**	**ASC**
Task Purpose Commander's intent	Task Purpose Commander's intent	Task Purpose Commander's intent

Constraints/issues/RFIs	**Host nation**
Highlight areas that need command emphasis	CBRNE asset support

Legend:

AO	area of operation
ASC	area support company
CBRNE	chemical, biological, radiological, nuclear, and explosives
Decon	decontamination
HRC	hazard response company
MOPP	mission-oriented protective posture
Recon	reconnaissance
RFI	request for information
TIM	toxic industrial material

Figure D-6. Example CBRN running staff estimate

D-16. Commanders determine the required protection for their units by assessing the capabilities of the enemy. They estimate the likely impact of CBRN attacks and, based on the concept of operations, determine the methods to reduce the impact and allow mission accomplishment. The CBRN vulnerability assessment considers the potential impact from a successful attack and the vulnerability of the force/facility/location to an attack. The vulnerability assessment provides insight into the unit ability to mitigate the risk. A miscalculation of the unit preparedness to execute CBRN defense has significant repercussions during execution.

D-17. Implementing many CBRN defensive measures may slow tempo, degrade combat power, and increase logistics requirements. CBRN R&S consumes resources, especially time. Personnel in protective equipment find it more difficult to work and fight. However, countering the CBRN threat with such measures is an essential component of preserving the force, assuring mobility, and protecting the scheme of maneuver against CBRN-related vulnerabilities.

D-18. Integrating CBRN defensive considerations into all types of rehearsals is central to mitigating CBRN-specific risk. As units prepare for execution, they must understand what they can and cannot do in MOPP4 gear. Movement and maneuver and fire support are harder to synchronize, rates of march can decrease significantly and lead to increased fuel usage, and many tactical tasks take longer to perform. Before initiating movement, commanders must have a thorough understanding of individual and collective proficiency in CBRN defense tasks. If units are untrained in these tasks, it may be necessary to generate branch plans to account for deficiencies. Taking time for the CBRN staff and leaders at echelon to address aspects of CBRN defense in combined arms and battle-drill rehearsals will reveal shortcomings and the required actions for mission success under CBRN conditions. This includes reviewing MOPP analysis and reinforcing MOPP guidance.

D-19. Considerations for commanders preparing for CBRN in decisive action include—

- Individual Soldier training in basic CBRN survival skills.
- The unit's proficiency in CBRN collective skills and mission-essential tasks under CBRN conditions.
- Subordinate leaders' understanding of CBRN hazards.
- The unit's ability to perform (project power) in MOPP4
- The unit's ability to continue the mission following a CBRN attack.

EXECUTION

D-20. Protection activities are continuous and enduring actions during execution. CBRN tasks and activities are conducted in offense, defense, and stability. Protection can be deliberately applied as commanders integrate and synchronize tasks and systems that comprise the protection warfighting function.

D-21. Units in MOPP 4 gear are at greater risk when executing tactical operations. With reduced situational awareness, disorientation, and difficulty communicating, units will operate in closer formations and take easier routes when maneuvering. As Soldiers remain in MOPP 4 for extended periods of time, the more their mission performance degrades due to increased dehydration, fatigue, inability to eat, and interference with bodily functions. CBRN reaction drills, not collectively understood or rehearsed in advance of chemical strikes, are ineffective and contribute to the generation of unnecessary casualties due to inadequate protection measures.

D-22. Donning the protective mask, alerting others of a CBRN strike, and wearing individual protective equipment are the most basic of tasks, yet Soldiers fail to train on them adequately. Units training as they fight, executing training at the platoon, company, and battalion levels in MOPP 4 will reduce the disruption normally experienced by conducting operations in IPE. Soldiers will become accustomed to communicating, managing work-rest cycles, and adjusting tempo appropriately before encountering a contaminated environment.

D-23. The full use of organic decontamination and detection systems must be enforced to the lowest echelon. Modified table of organization and equipment authorizations include detection systems at platoon level and above, capable of identifying chemical warfare agents and radiation. Platoon level detection equipment, employed by trained Soldiers in advance of a chemical strike, is a valuable tool for early warning and identification. Battalions organically possess the assets to conduct operational decontamination for their subordinate units. They must design their utilization of those resources to form an organized capability, and then develop training plans to sustain the combat power of that element. Unit decontamination equipment, with dedicated teams, provides the freedom and flexibility for battalions to execute decontamination on their terms. Battalions will not have to depend on external enablers for decontamination support. Investing in this capability will significantly reduce wait times for decontamination and allow rapid regeneration of combat power.

D-24. Improving operational decontamination capability starts by continually verifying the maintenance status of decontamination equipment and implementing a battalion training program for operational decontamination teams. Formalizing this requirement with emphasis from battalion leadership and identifying specific individuals to execute operational decontamination (usually from the headquarters and headquarters company) are often eclipsed by other training requirements.

ASSESSMENT

D-25. Leaders and staffs conduct continuous assessments of CBRN activates throughout the operations process, regardless of the echelon of command, OE, or operational phase. Protection assessment is an essential, continuing activity that occurs throughout the operations process. Activities include maintaining situational understanding, monitoring and evaluating running estimates. The protection working group continually assess threats to protection priorities.

OTHER STAFF FUNCTIONS

D-26. Antiterrorism and force protection is extremely important given the ever present threat of terrorist attacks, insider threats, and the need to protect our forces from becoming targets of opportunity. The antiterrorism program is a collective, proactive effort, focused on the prevention and detection of terrorist attacks against DOD personnel and their families, facilities, installations, and infrastructure critical to mission accomplishment as well as the preparation to defend against and planning for the response to the consequences of terrorist incidents. antiterrorism planning is integrated into the overall force protection planning as prescribed in DODI 2000.12.

D-27. CBRN staff coordinates and synchronizes terrorism-focused risk assessments with CBRN risk management requirements prescribed in DODI 3150.09. As the force protection condition level changes, protection against CBRN threats and hazards is accomplished through an interoperable system of surveillance, monitoring, and response. Therefore, CBRN staff must be familiar with DOD force protection condition levels prescribed in DODI O-2000.16, volume 2 to ensure an appropriate measure of protection against CBRN effects is addressed for each level. CBRN staff will serve as the subject matter expert for the commander to assess compliance, effectiveness, and adequacy of subordinate organizations and provide recommendations to enhance the overall antiterrorism/force protection program. This comprehensive assessment should not be misinterpreted to be an inspection.

CBRN DEFENSE INTEGRATION, COLLABORATION, AND SYNCHRONIZATION

D-28. Exclusive use of organizational CBRN sensors limits hazard awareness and understanding. Integration across the spectrum of sensors (CBRN and non-CBRN) provides a more effective understanding of the CBRN OE. It is only through the integration of capabilities across the warfighting functions that situational understanding can be achieved. When operating in a CBRN environment, the tactical commander is challenged to make timely and informed risk based decisions, supported by awareness and understanding, in order to ensure survivability while retaining freedom of action in accomplishing the mission.

D-29. The integration, collaboration, and synchronization of the warfighting functions primarily occurs within the staffs. Integration of all-source intelligence and information incorporating CBRN sensing and detection capabilities provides a better picture of the OE and facilitation of situational understanding. The other staff sections coordinate and integrate the requirements of the CBRN staff. More considerations for the warfighting functions can be found in appendix B.

- Assistant chief of staff, personnel (G-1)/battalion or brigade personnel staff officer (S-1). The CBRN staff collaborates with the personnel section for status of CBRN personnel, CBRN personnel assignments and by using advanced hazard prediction modeling for predicting CBRN casualties. The chaplain section (echelons above brigade) or unit ministry team (brigade and battalion) should be collaborated with for provision of religious support to Soldiers, DOD Civilians, and authorized civilians as well as advisement on religion, ethics, and morale following a CBRN incident.
- G-2/S-2. The CBRN staff works with intelligence section to understand enemy CBRN capabilities and vulnerabilities, current and projected weather, and intent to use CBRN weapons. Information obtained by intelligence helps define threat characteristics, detailed terrain information and weather. The G-9 and associated Civil Affairs forces can provide the most up-to-date information on civil considerations. All of this information feeds into the CBRN staff's ability to prepare accurate threat assessments. The products of the CBRN staff analysis feed into IPB and other staff sections mission analysis results.

- G-3/battalion or brigade operations staff officer (S-3) (operations). The CBRN staff provides information on CBRN defense training, vulnerabilities, effects on operations, employment of CBRN units.
- Assistant chief of staff, logistics (G-4)/battalion or brigade logistics staff officer (S-4) (logistics). The CBRN staff coordinates with logistics staffs to plan CBRN defense equipment and supplies requirements, maintenance of CBRN equipment and transportation of CBRN assets.
- Assistant chief of staff, plans (G-5). The CBRN staff coordinates with the G-5 for information on host-nation forces decontamination and reconnaissance that are locally available to support, train, and equip local nationals for CBRN defense.
- G-9. The Civil Affairs operations staff officer is responsible for the planning, integrating, evaluating, and assessment of civil considerations into the MDMP and Army design methodologies. Civil Affairs operations establish liaisons with civilian organizations to coordinate their efforts with Army CBRN defense and CWMD operations. These efforts protect indigenous populations and institutions and limit collateral damage.
- CBRN staff may use automated hazard prediction modeling capabilities to estimate civilian casualties and to model effects on civilian populations.
- Command surgeon and medical staff. The command surgeon and medical staff should participate in the CBRN and intelligence staffs' working groups and meetings in order to—
    - Advise on medical CBRN defensive actions for related health threats (such as immunizations and pretreatments, protection against diseases and contaminated food and water).
    - Advise on chemical, biological, and radiological related health status of U.S., multinational, and enemy forces.
    - Coordinate laboratory analysis of CBRN environmental samples and clinical specimens.
    - Ensure documentation of potential or known Soldier CBRN hazard exposures in order to inform the commanders of potential signs and symptoms, mission degradation, and need for increase medical surveillance and possible CBRN hazard specific treatment.

Appendix E
Training

ESSENTIAL CBRN PROFICIENCY SKILLS

E-1. Training for officers, enlisted personnel, and civilian support personnel whose primary duties are concerned with the planning, coordination, supervision, and conduct of unit CBRN defense activities should include essential CBRN proficiency skills. These personnel should receive formal training that meets the requirements consistent with those duties. Table E-1 identifies individual tasks for CBRN defense specialists, to include unit CBRN defense officers, enlisted personnel, and DOD civilians (including private contractors) assigned on an additional-duty basis to form the CBRN control party. These personnel should be at the company, battery, or troop task unit, but they may be at a higher level, depending on the organizational structure.

Table E-1. Individual tasks for CBRN defense specialists

Task number	*Title*
031-74D-1010	Collect a Solid Chemical Sample
031-74D-1014	Identify Decontamination Methods
031-74D-1015	Conduct Operator Wipe Down Using the M100 Sorbent Decontamination System (SDS)
031-74D-1016	Detect Chemical Agent Vapors Using the Improved Chemical-Agent Monitor (ICAM)
031-74D-1017	Emplace a Chemical Agent Alarm
031-74D-1018	Collect a Liquid Chemical Sample
031-74D-1019	Detect Chemical Warfare Agents Vapors Using the Joint Chemical Agent Detector (JCAD)
031-74D-1020	Decontaminate Equipment Using the M26 Decontamination Apparatus
031-74D-2011	Submit RADIAC Equipment to Test, Measurement and Diagnostic Equipment (TMDE)
031-74D-2012	Manage Operational Decontamination
031-74D-2013	Establish a Detailed Troop Decontamination Line
031-74D-2014	Process Soldiers Through a Detailed Troop Decontamination
031-74D-2016	Establish a Detailed Equipment Decontamination Line
031-74D-2017	Process Vehicles Through a Detailed Equipment Decontamination
031-74D-2018	Plan Thorough Decontamination
031-74D-3000	Select Domestic CBRN Incident Decontamination Site
031-74D-3012	Plan Operational Decontamination
031-74D-6000	Plan for CBRN Defense Equipment in Support of a Deployment
Legend:	
CBRN	chemical, biological, radiological and nuclear
RADIAC	radioactivity detection indication and computation

COMMON CBRN DEFENSE INDIVIDUAL TASKS FOR ALL UNITS

E-2. All Service personnel must be trained in the basic concepts of CBRN defense to survive a CBRN-related attack and contribute to the survivability and operating proficiency of the organization in a CBRN environment. Individual standards of proficiency include common tasks that individuals must master to survive a CBRN attack. Table E-2 identifies common tasks for CBRN defense. These are nonmilitary occupational specialty specific individual tasks.

Table E-2. Common tasks for CBRN defense

<table>
<tr><th>Task number</th><th colspan="2">Title</th></tr>
<tr><td>031-COM-1000</td><td colspan="2">Conduct MOPP Gear Exchange</td></tr>
<tr><td>031-COM-1001</td><td colspan="2">React to a Nuclear Attack</td></tr>
<tr><td>031-COM-1002</td><td colspan="2">React to Depleted Uranium</td></tr>
<tr><td>031-COM-1003</td><td colspan="2">Mark CBRN-Contaminated Areas</td></tr>
<tr><td>031-COM-1004</td><td colspan="2">Protect Yourself From Chemical and Biological Contamination Using Your Assigned Protective Mask</td></tr>
<tr><td>031-COM-1005</td><td colspan="2">Protect Yourself From CBRN Injury/Contamination by Assuming MOPP Level 4</td></tr>
<tr><td>031-COM-1006</td><td>Decontaminate Your Skin</td><td rowspan="2">In STP 21-1 SMCT, these are one combined task: 031-COM-1006</td></tr>
<tr><td>031-COM-1011</td><td>Decontaminate Individual Equipment</td></tr>
<tr><td>031-COM-1007</td><td colspan="2">React to Chemical or Biological Hazard/Attack</td></tr>
<tr><td>031-COM-1008</td><td colspan="2">Identify Liquid Chemical Agents Using M8 Paper</td></tr>
<tr><td>031-COM-1009</td><td colspan="2">Detect Liquid Chemical Agents Using M9 Detector Paper</td></tr>
<tr><td>031-COM-1010</td><td colspan="2">Maintain Your Assigned Protective Mask</td></tr>
<tr><td>031-COM-1012</td><td colspan="2">Conduct Personal Hydration While Wearing Your Assigned Protective Mask</td></tr>
<tr><td>031-COM-2000</td><td colspan="2">Conduct Unmasking Procedures</td></tr>
<tr><td>031-COM-2001</td><td colspan="2">Report a CBRN Attack Using a CBRN 1 Report</td></tr>
<tr><td>031-COM-2004</td><td colspan="2">Identify Chemical Agents Using a M256A2 Chemical-Agent Detector Kit</td></tr>
<tr><td colspan="3">Legend:
CBRN chemical, biological, radiological and nuclear
MOPP mission-oriented protective posture
SMCT Soldier's manual of common tasks
STP Soldier training publication</td></tr>
</table>

COLLECTIVE TASKS FOR UNITS CONDUCTING OPERATIONS IN CBRN ENVIRONMENTS

E-3. Effective mission command philosophy is conducive to setting the conditions for battle-focused training and for the development of effective teams. Due to reporting and synchronization requirements, CBRN incidents incorporated into decisive action provide a multiechelon training venue that emphasizes mission command. Although staffs and crews train battle drills individually on reacting to CBRN incidents, battle-focused training emphasizes that planners should replicate the conditions of a CBRN incident from initial observation, warning and reporting, synchronization between higher and adjacent units, and final resolution.

E-4. Commanders use unit training plans to address CBRN threats as a complex variable of the OE that directly impacts a unit's ability to execute mission-essential tasks. Unit NCOs identify CBRN supporting collective tasks, by echelon, to mitigate CBRN impact on a unit mission. The following recommended collective tasks provide commanders with a training strategy to achieve proficiency in a contested CBRN environment. Tables E-3 through E-8, pages E-3 through E-6, describe the recommended collective tasks for each echelon.

Table E-3. Recommended collective tasks for a division

Task Number	*Task*
03-DIV-0065	Prepare for Operations Under CBRN Conditions
03-DIV-0066	Prepare for Nuclear Attack
03-DIV-0067	Prepare for a Friendly Nuclear Strike
03-DIV-0069	React to a Nuclear Attack
03-DIV-0070	Prepare for a Chemical Attack
03-DIV-0071	React to a Chemical Agent Attack
03-DIV-0404	Direct CBRN Defense Operations
Legend: CBRN	 chemical, biological, radiological, and nuclear

Table E-4. Recommended collective tasks for a brigade

Task Number	*Task*
03-BDE-0065	Prepare for Operations Under CBRN Conditions
03-BDE-0066	Prepare for Nuclear Attack
03-BDE-0067	Prepare for a Friendly Nuclear Strike
03-BDE-0069	React to a Nuclear Attack
03-BDE-0070	Prepare for a Chemical Attack
03-BDE-0071	React to a Chemical Agent Attack
Legend: CBRN	 chemical, biological, radiological, and nuclear

Table E-5. Recommended collective tasks for a battalion

Task Number	*Task*
03-BN-0065	Prepare for Operations Order CBRN Conditions
03-BN-0066	Prepare for Nuclear Attack
03-BN-0067	Prepare for a Friendly Nuclear Strike
03-BN-0069	React to a Nuclear Attack
03-BN-0070	Prepare for a Chemical Attack
03-BN-0071	React to a Chemical Agent Attack
Legend: CBRN	 chemical, biological, radiological, and nuclear

Table E-6. Recommended collective tasks for a company/battery/troop

Task Number	*Task*
03-CO-9200	Conduct Unmasking Procedures After a Chemical Attack
03-CO-9201	Implement CBRN Protective Measures
03-CO-9203	React to a Chemical Biological (CB) Attack
03-CO-9204	Conduct Actions After a Chemical Biological (CB) Attack
03-CO-9205	Prepare for a Friendly Nuclear Strike
03-CO-9208	Cross a Radiological Contaminated Area
03-CO-9209	React to Obscuration
03-CO-9223	React to a Nuclear Attack

Table E-6. Recommended collective tasks for a company/battery/troop (continued)

03-CO-9224	Conduct Operational Decontamination
03-CO-9225	Conduct a Chemical Reconnaissance
03-CO-9226	Cross a Chemically Contaminated Area
03-CO-9310	Conduct a Dismounted Chemical Survey
03-CO-9312	Conduct Thorough Decontamination

Legend:
CBRN chemical, biological, radiological, and nuclear

Table E-7. Recommended collective tasks for organic reconnaissance platoons to a BCT

Task Number	*Task*	*ABCT*	*IBCT*	*SBCT*	*Comments*
03-PLT-0001	Conduct CBRN Mounted Reconnaissance Locate	X		X	CBRN Recon PLT
03-PLT-0002	Conduct CBRN Mounted Reconnaissance Survey	X		X	CBRN Recon PLT
03-PLT-0003	Conduct CBRN Mounted Surveillance	X		X	CBRN Recon PLT
03-PLT-0004	Conduct CBRN Mounted Reconnaissance Sampling	X		X	CBRN Recon PLT
03-PLT-0005	Conduct Site Characterization		X		CBRN Recon PLT
03-PLT-0008	Plan a CBRN Reconnaissance and Surveillance (R&S) Mission	X	X	X	CBRN Recon PLT
03-PLT-0009	Prepare for a Mission	X	X	X	CBRN Recon PLT
03-PLT-0019	Identify Biological Surveillance Sites	X		X	CBRN Recon PLT
03-PLT-0022	Conduct Biological Surveillance	X		X	CBRN Recon PLT
03-PLT-0028	Prepare a Biological Sample for Evacuation	X		X	CBRN Recon PLT
03-PLT-0029	Evacuate Biological Samples to the Designated Sample Transfer Point	X		X	CBRN Recon PLT
03-PLT-0030	Coordinate with the Unit Commander or Higher Headquarters (HQ) for Unit Employment	X		X	CBRN Recon PLT
03-PLT-0031	Emplace a Biological Detection System	X		X	CBRN Recon PLT
03-PLT-0032	Displace a Biological Detection System	X		X	CBRN Recon PLT
03-PLT-0038	Conduct Biological Detection (BD) Data Analysis	X		X	CBRN Recon PLT
03-PLT-0044	Conduct CBRN Dismounted Reconnaissance Locate		X		CBRN Recon PLT
03-PLT-0045	Conduct CBRN Dismounted Reconnaissance Survey		X		CBRN Recon PLT
03-PLT-0046	Conduct CBRN Dismounted Surveillance		X		CBRN Recon PLT
03-PLT-0047	Conduct CBRN Dismounted Reconnaissance Sampling		X		CBRN Recon PLT
03-PLT-0048	Conduct Aerial CBRN Reconnaissance	X	X	X	CBRN Recon PLT with aircraft support
03-PLT-1076	Conduct a Toxic Industrial Material Reconnaissance	X	X	X	CBRN Recon PLT would be used if industrial facility was destroyed
03-PLT-5129	Conduct Technical Decontamination		X		CBRN Recon PLT
03-PLT-9225	Conduct a Chemical Reconnaissance		X		CBRN Recon PLT

Legend:
ABCT armored brigade combat team
CBRN chemical, biological, radiological, and nuclear
IBCT infantry brigade combat team
PLT platoon
Recon reconnaissance
SBCT Stryker brigade combat team

Table E-8. Recommended collective tasks for staffs

Task Number	*Task*	*BN*	*BDE/GRP*	*DIV*	*CORP*	*EAC*	*MEB*	*ABCT*	*IBCT*	*SBCT*	*Comments*
03-SEC-0002	Maintain a CBRN Running Estimate	X	X	X	X	X	X	X	X	X	Staff task
03-SEC-0003	Maintain the CBRN Common Operating Picture (COP)	X	X	X	X	X	X	X	X	X	Staff task
03-SEC-0004	Plan Chemical-Unit Employment	X	X	X			X	X	X	X	Staff task
03-SEC-0006	Prepare a Toxic Industrial Material (TIM) Release Vulnerability Analysis	X	X	X	X	X	X	X	X	X	Staff task
03-SEC-0007	Prepare a Chemical Vulnerability Analysis	X	X	X	X	X	X	X	X	X	Staff task
03-SEC-0008	Prepare a Biological Vulnerability Analysis	X	X	X	X	X	X	X	X	X	Staff task
03-SEC-0009	Prepare a Nuclear Vulnerability Analysis	X	X	X	X	X	X	X	X	X	Staff task
03-SEC-0010	Perform Vulnerability Assessment	X	X	X	X	X	X	X	X	X	Staff task
03-SEC-0011	Process Weather Data	X	X	X	X	X	X	X	X	X	Staff task
03-SEC-0012	Prepare Contamination Predictions	X	X	X	X	X	X	X	X	X	Staff task
03-SEC-0013	Coordinate Chemical/Biological Survey/Sampling Operations	X	X	X			X	X	X	X	Staff task
03-SEC-0014	Coordinate Radiological Survey Missions	X	X	X			X	X	X	X	Staff task
03-SEC-0015	Recommend Operational Exposure Guidance	X	X	X	X	X	X	X	X	X	Staff task
03-SEC-0017	Prepare Appendix 10 (CBRN Defense) to Annex E (Protection)	X	X	X	X	X	X	X	X	X	Staff task
03-SEC-0018	Prepare for a Biological Attack	X	X	X	X	X	X	X	X	X	Staff task
03-SEC-0019	React to a Biological Attack	X	X	X	X	X	X	X	X	X	Staff task
03-SEC-0020	Perform Operational Environment Assessment	X	X	X	X	X	X	X	X	X	Staff task
03-SEC-0080	Plan Biological Surveillance and Sampling Operations	X	X	X			X	X	X	X	Staff task
03-SEC-1017	Monitor CBRN and Obscuration Missions	X	X	X			X	X	X	X	Staff task
03-SEC-1018	Conduct CBRN Reconnaissance and Surveillance (R&S) Planning	X	X	X			X	X	X	X	Staff task
03-SEC-1020	Prepare a CBRN Information Collection Plan	X	X	X	X	X	X	X	X	X	Staff task
03-SEC-1810	Plan for the Employment of Obscurants	X	X	X	X	X	X	X	X	X	Staff task
03-SEC-1811	Prepare Appendix 9 (Battlefield Obscuration) to Annex C (Operations)	X	X	X	X	X	X	X	X	X	Staff task
03-SEC-9002	Develop a CBRN Defense Plan	X	X	X	X	X	X	X	X	X	Staff task
03-SEC-9007	Coordinate CBRN Protection	X	X	X	X	X	X	X	X	X	Staff task
03-SEC-9012	Process CBRN Reports	X	X	X	X	X	X	X	X	X	Staff task
03-SEC-9020	Implement the CBRN Warning and Reporting System			X	X	X					Staff task
03-SEC-9791	Facilitate CBRN response operations		X	X	X	X	X	X	X	X	Staff task

Table E-8. Recommended collective tasks for staffs (continued)

Legend:			
ABCT	armored brigade combat team	EAC	echelon above corps
BDE	brigade	GRP	group
BN	battalion	IBCT	infantry brigade combat team
CBRN	chemical, biological, radiological, and nuclear	MEB	maneuver enhancement brigade
DIV	division	SBCT	Stryker brigade combat team

ADDITIONAL PROFICIENCY SKILLS FOR MEDICAL PERSONNEL

E-5. Medical personnel protect themselves, patients, and medical facilities against exposure to CBRN agents (CBRN defense) and, according to the latest developments in science and technology, carry out the measures necessary to maintain and restore the health of personnel exposed to CBRN environments (CBRN medical defense).

E-6. Table E-9 identifies additional proficiency skills that are required for medical personnel. These are in addition to the skills outlined in previous paragraphs that apply according to their rank and function.

Table E-9. Additional proficiency skills for medical personnel

Additional CBRN Skills for Medical Personnel
Effectively protect casualties in a CBRN scenario during first aid, triage, resuscitative and emergency treatment, evacuation, and hospital treatment.
Consider actions that provide an optimum level of protection against CBRN hazards for medical materials, vehicles, and facilities.
Be familiar with fielded collective protection systems for facilities and vehicles, if appropriate.
Possess good knowledge of the acute symptoms of CBRN injuries and their specific countermeasures and potential side effects.
Possess good knowledge of decontamination procedures for CBRN-contaminated patients.
Additional Skills for Trained Medical Officers and Noncommissioned Officers
Have specialized knowledge in contamination control procedures for CBRN-contaminated patients and associated equipment (including RADIAC monitors and chemical-agent monitors). (Applicable to selected medical personnel.)
Have task-oriented, specialized knowledge of the diagnosis and treatment of CBRN injuries and the detection and identification of chemical and biological agents and radiation. (Applicable to medical personnel assigned to perform special CBRN medical defense tasks during missions [anesthesiologists, surgeons, internists, microbiologists, food protection personnel, veterinarians]).
Be able to convert scientific expert reports into clear advice to the commander. (Applicable to staff and command surgeons.)
Have knowledge of the acute and long-term health effects of CBRN hazards in the deployment area and of the follow-on medical support requirements from those hazards. (Applicable to selected medical personnel with advanced training.)
Have knowledge of the risk benefit balance from wearing individual protective equipment and using prophylactic medical CBRN countermeasures. (Applicable to selected medical personnel with advanced training.)

Table E-9. Additional proficiency skills for medical personnel (continued)

Action Before, During, and After an Operation
Establish an inventory of CBRN hazards and infectious endemic diseases in the deployment area, and establish the resulting medical support requirements in relation to countermeasures.
Document and register the position of personnel during possible exposures and the level of protection from that exposure.
Coordinate investigations of unusual sickness and fatalities in situations involving CBRN hazards and endemic diseases.
Conduct outbreak management in the case of highly contagious diseases in a biological scenario.
Conduct postconflict surveillance for illnesses and follow up exposed or potentially exposed forces.
Legend: CBRN chemical, biological, radiological and nuclear RADIAC radioactivity detection indication and computation

LINKAGE TO SUPPORTING DOCTRINE

E-7. The techniques nested within this publication provide the methods for fulfilling CBRN functions. CBRN operations tasks can be found in supporting doctrine, as depicted in figure E-1, page E-8. The following bullets describe these tasks and their corresponding reference publications:

- Tasks such as conducting threat assessments, advising the commander on CBRN operations, and conducting reconnaissance and surveillance associated with assessing CBRN hazards may be found in—
  - ATP 3-11.36.
  - ATP 3-11.37.
  - TM 3-11.91.
- Tasks for CBRN defense that are associated with protecting the force in CBRN environments can be found in ATP 3-11.32.
- Tasks to mitigate (decontamination, small-scale elimination, domestic CBRN response) can be found in—
  - ATP 3-11.23.
  - ATP 3-11.32.
  - ATP 3-11.41.
  - TM 3-11.91.
- Tasks associated with the integrating activity of hazard awareness and understanding can be found in ATP 3-11.36.
- The doctrinal collective tasks associated with this publication can be found in ADRP 1-03.

FM 3-11 Intent:

Provide commanders and staffs with overarching doctrine for operations in CBRN environments.

Establish the functions of assess threats and hazards, provide protection, and mitigate CBRN incidents

JP 3-11
JP 3-40
JP 3-41
ADP 3-0
ADP 3-37

FM 3-0
FM 3-11 (2011)

FM 3-11 (Revised)

ATP 3-11.23, MTTP for *WMD-E Operations*

ATP 3-11.24, *Technical CBRNE Force Employment*

ATP 3-11.32, MTTP *CBRN Passive Defense*

ATP 3-11.36, MTTP *CBRN Planning*

ATP 3-11.37, MTTP *CBRN Reconnaissance and Surveillance*

ATP 3-11.41, *CBRN Consequence Management*

ATP 3-11.46, *WMD-CST Operations*

ATP 3-11.47, *CERFP/HRF Operations*

ATP 3-37.11, *CBRNE Command*

TM 3-11.32, MTTP *CBRN Warning and Reporting and Hazard Prediction Procedures*

TM 3-11.91, *CBRN Threats and Hazards*

Notes:

FM 3-11 ties doctrine publications such as ADP 3-37, ADPs 3-0, 5-0, and 6-0 to CBRN operations tactics descriptively for the maneuver commander.

Techniques for CBRN Soldiers and those who perform CBRN functions will be found in appropriate Army techniques publications.

Legend:

ADP	Army doctrine publication
ATP	Army techniques publication
CBRN	chemical, biological, radiological, and nuclear
CBRNE	chemical, biological, radiological, and nuclear explosives
CERFP-HRT	chemical, biological, radiological, nuclear, and high-yield explosives enhanced response force package/homeland response force
FM	field manual
JP	joint publication
MTTP	multi-Service tactics, techniques, and procedures
TM	technical publication
WMD-CST	weapons of mass destruction–civil support team
WMD-E	weapons of mass destruction–elimination

Figure E-1. FM 3-11 to ATP transition model

Glossary

The glossary lists acronyms and terms with Army or joint definitions. Where Army and joint definitions differ, (Army) precedes the definition. Terms for which FM 3-11 is the proponent manual are marked with an asterisk (*). The proponent manual for other terms is listed in parentheses after the definition.

SECTION I – ACRONYMS AND ABBREVIATIONS

AA	assembly area
ABCT	armored brigade combat team
ADP	Army doctrine publication
ADRP	Army doctrine reference publication
AFTTP	Air Force tactics, techniques, and procedures
AO	area of operations
ARSOF	Army Special Operations Forces
ASCOPE	areas, structures, capabilities, organizations, people, and events
attn	attention
ATP	Army techniques publication
BCT	brigade combat team
BEB	brigade engineer battalion
CBRN	chemical, biological, radiological, and nuclear
CBRNE	chemical, biological, radiological, nuclear, and explosives
CBRNWRS	chemical, biological, radiological, and nuclear warning and reporting system
CCIR	commander's critical information requirement
CDID	Capabilities Development and Integration Directorate
COA	course of action
COCOM	combatant command
CODDD	Concepts, Organizations, and Doctrine Development Division
COLPRO	collective protection
CS	chlorobenzylidenemalononitrile
CWMD	countering weapons of mass destruction
DA	Department of the Army
DOD	Department of Defense
DODD	Department of Defense directive
DODI	Department of Defense instruction
DSCA	defense support of civil authorities
DTRA	Defense Threat Reduction Agency
EA	engagement area
EOD	explosive ordnance disposal

FM	field manual
FPCON	force protection condition
G-1	assistant chief of staff, personnel
G-2	assistant chief of staff, intelligence
G-3	assistant chief of staff, operations
G-4	assistant chief of staff, logistics
G-5	assistant chief of staff, plans
G-9	assistant chief of staff, civil affairs operations
GTA	graphic training aid
IBCT	infantry brigade combat team
ICBRN-R	international chemical, biological, radiological, and nuclear response
IPB	intelligence preparation of the battlefield
JP	joint publication
MCRP	Marine Corps reference publication
MCWP	Marine Corps warfighting publication
MDMP	military decisionmaking process
MEB	maneuver enchancement brigade
METT-TC	mission, enemy, terrain and weather, troops and support available, time available, and civil considerations
MO	Missouri
MOPP	mission-oriented protective posture
MSCoE	Maneuver Support Center of Excellence
MSR	main supply route
NATO	North Atlantic Treaty Organization
NGIC	National Ground Intelligence Center
NTRP	Navy tactical reference publication
NTTP	Navy tactics, techniques, and procedures
OAKOC	observation and fields of fire, avenues of approach, key terrain, obstacles, and cover and concealment
OBJ	objective
ODIN	OE Data Integration Network
OE	operational environment
PAA	position area for artillery
PIR	priority intelligence requirement
PL	phase line
PMESSI-PT	political, military, economic, social, information, infrastructure, physical environment, and time
R&S	reconnaissance and surveillance
ROMO	range of military operations
S-1	battalion or brigade personnel staff officer
S-2	battalion or brigade intelligence staff officer
S-3	battalion or brigade operations staff officer

S-4	battalion or brigade logistics staff officer
S-9	battalion or brigade civil affairs operations staff officer
SBF	support by fire
STANAG	standardization agreement
STP	Soldier training publication
SWEAT-MSO	sewage, water, electricity, academics, trash, medical, safety, other considerations
TC	training circular
TIB	toxic industrial biological
TIM	toxic industrial material
TM	technical manual
TRADOC	United States Army and Training Doctrine Command
U.S.	United States
USACBRNS	United States Army Chemical, Biological, Radiological, and Nuclear School
USANCA	United States Army Nuclear and Countering Weapons of Mass Destruction Agency
WMD	weapons of mass destruction
WWI	World War I

SECTION II – TERMS

***chemical, biological, radiological, and nuclear operations**

Chemical, biological, radiological, and nuclear operations include the employment of capabilities that assess, protect against, and mitigate the entire range of chemical, biological, radiological, and nuclear incidents to enable freedom of action.

chemical, biological, radiological, nuclear, and explosives

Components that are threats or potential hazards with adverse effects in the operational environment. (ATP 3-37.11)

consolidate gains

Activities to make enduring any temporary operational success and set the conditions for a stable environment allowing for a transition of control to legitimate authorities. (ADP 3-0)

contamination mitigation

The planning and actions taken to prepare for, respond to, and recover from contamination associated with all chemical, biological, radiological, and nuclear threats and hazards to continue military operations. (JP 3-11)

countering weapons of mass destruction

Efforts against actors of concern to curtail the conceptualization, development, possession, proliferation, use, and effects of weapons of mass destruction, related expertise, materials, technologies, and means of delivery. (JP 3-40)

hazard

A condition with the potential to cause injury, illness, or death of personnel; damage to or loss of equipment or property; or mission degradation. (JP 3-33)

protection

Preservation of the effectiveness and survivability of mission-related military and nonmilitary personnel, equipment, facilities, information, and infrastructure deployed or located within or outside the boundaries of a given operational area. (JP 3-0)

threat

Any combination of actors, entities, or forces that have the capability and intent to harm United States forces, United States national interests, or the homeland. (ADP 3-0)

weapons of mass destruction

Weapons of mass destruction are chemical, biological, radiological, or nuclear weapons capable of causing a high order of destruction or causing mass casualties and excluding the means of transporting or propelling the weapon where such means is a seaparable and divisible from the weapon. (JP 3-40)

References

All URLs were accessed 11 December 2018.

REQUIRED PUBLICATIONS

These documents must be available to the intended users of this publication. Most Army publications are available online at <https://armypubs.army.mil>. Most joint publications are available online at <http://www.jcs.mil/Doctrine/>.

DOD Dictionary. *DOD Dictionary of Military and Associated Terms.* April 2019.

ADP 1-02. *Terms and Military Symbols.* 14 August 2018.

RELATED PUBLICATIONS

These documents contain relevant supplement information.

JOINT PUBLICATIONS

Most joint publications are available online at <www.jcs.mil/doctrine/>.

JP 3-0. *Joint Operations*. 17 January 2017.

JP 3-11. *Operations in Chemical, Biological, Radiological, and Nuclear Environments.* 29 October 2018.

JP 3-17. *Air Mobility Operations*. 5 February 2019.

JP 3-33. *Joint Task Force Headquarters*. 31 January 2018.

JP 3-40. *Countering Weapons of Mass Destruction.* 31 October 2014.

JP 3-41. *Chemical, Biological, Radiological, and Nuclear Response*. 9 September 2016.

JP 4-0. *Joint Logistics.* 4 February 2019.

ARMY PUBLICATIONS

Most Army publications are available online at <https://armypubs.army.mil>.

ADP 1-01. *Doctrine Primer.* 2 September 2014.

ADP 3-0. *Operations.* 6 October 2017.

ADP 3-28. *Defense Support of Civil Authorities.* 11 February 2019.

ADP 3-37. *Protection.* 11 December 2018.

ADP 3-90. *Offense and Defense.*13 August 2018.

ADP 5-0. *The Operations Process.* 17 May 2012.

ADP 6-0. *Mission Command.* 17 May 2012.

ADRP 1-03. *The Army Universal Task List.* 2 October 2015.

ADRP 3-0. *Operations.* 6 October 2017.

ADRP 3-05. *Special Operations.* 29 January 2018.

ATP 2-01.3. *Intelligence Preparation of the Battlefield.* 1 March 2019.

ATP 3-05.11. *Special Operations Chemical, Biological, Radiological, and Nuclear Operations.* 30 April 2014.

ATP 3-11.24. *Technical Chemical, Biological, Radiological, Nuclear, and Explosives Force Employment.* 6 May 2014.

ATP 3-11.46. *Weapons of Mass Destruction—Civil Support Team Operations.* 20 May 2014.

ATP 3-11.47. *Chemical, Biological, Radiological, Nuclear, and High-Yield Explosives Enhanced Response Force Package (CERFP)/Homeland Response Force (HRF) Operations.* 26 April 2013.

ATP 3-21.51. *Subterranean Operations.* 21 February 2018.

ATP 3-37.11. *Chemical, Biological, Radiological, Nuclear, and Explosives Command.* 28 August 2018.

ATP 3-60. *Targeting.* 7 May 2015.

ATP 3-90.40. *Combined Arms Countering Weapons of Mass Destruction.* 29 June 2017.

ATP 4-32. *Explosive Ordnance Disposal (EOD) Operations.* 30 September 2013.

ATP 4-46. *Contingency Fatality Operations.* 17 December 2014.

FM 3-0. *Operations.* 6 October 2017.

FM 3-55. *Information Collection.* 3 May 2013.

FM 3-57. *Civil Affairs Operations.* 17 April 2019.

FM 3-81. *Maneuver Enhancement Brigade.* 21 April 2014.

FM 3-90-1. *Offense and Defense Volume 1.* 22 March 2013.

FM 3-94. *Theater Army, Corps, and Division Operations.* 21 April 2014.

FM 6-0. *Commander and Staff Organization and Operations.* 5 May 2014.

FM 6-99. *U.S. Army Report and Message Formats.* 19 August 2013.

FM 7-0. *Train to Win in a Complex World.* 5 October 2016.

FM 27-10. *The Law of Land Warfare.* 18 July 1956.

GTA 03-06-008. *CBRN Warning and Reporting System.* June 2017.

STP 21-1-SMCT. *Soldier's Manual of Common Tasks, Warrior Skills Level 1.* 28 September 2017.

STP 21-24-SMCT. *Soldier's Manual of Common Tasks Warrior Leader Skills Level 2, 3, and 4.* 9 September 2008.

TC 7-100.2. *Opposing Force Tactics.* 9 December 2011.

Department of Defense Publications

Most joint and Department of Defense publications are available online at <http://www.esd.whs.mil/DD/>.

DODD 3025.18. Defense Support of Civil Authorities (DSCA).29 December 2010. Web site at <https://www.esd.whs.mil/Directives/issuances/dodd/>, accessed 15 May 2019.

DODI 2000.12. *DOD Antiterrorism (AT) Program.* 1 March 2012. Web site at <http://www.esd.whs.mil/Directives/issuances/dodi/>, accessed 17 October 2018.

DODI O-2000.16, Volume 2. *DOD Antiterrorism (AT) Program Implementation: DOD Force Protection Condition (FPCON) System.* 17 November 2016. Web site at <http://www.esd.whs.mil/Directives/issuances/dodi/>, accessed 17 October 2018.

DODI 3150.09. *The Chemical, Biological, Radiological, and Nuclear (CBRN) Survivability Policy.* 8 April 2015. Web site at <http://www.esd.whs.mil/Directives/issuances/dodi/>, accessed 17 October 2018.

Miscellaneous Publications

National Defense Strategy 2018. 19 January 2018. Web site <https://dod.defense.gov/Portals/1/Documents/pubs/2018-National-Defense-Strategy-Summary.pdf>, accessed 15 May 2019.

National Military Strategy 2018. December 2018. Web site <http://nssarchive.us/national-military-strategy-2018/>, accessed 15 May 2019.

National Security Strategy 2017. <https://www.hsdl.org/?abstract&did=806478>, accessed 12 December 2018.

National Strategy for Countering Weapons of Mass Destruction Terrorism. December 2018. Web site <https://www.whitehouse.gov/wp-content/uploads/2018/12/20181210_National-Strategy-for-Countering-WMD-Terrorism.pdf>, accessed 15 May 2019.

ODIN Worldwide Equipment Guide. <https://odin.tradoc.army.mil>, accessed 17 October 2018.

STANAG 2103 (ATP-45). *Warning and Reporting and Hazard Prediction of Chemical, Biological, Radiological, and Nuclear Incidents (Operator's Manual), Edition 11 (ATP-45, Edition E, Version 1)*. 23 January 2014. <https://nso.nato.int/protected/nsdd/stanagdetails.html?idCover=8467&LA=EN>, accessed 17 October 2018.

MULTI-SERVICE PUBLICATIONS

ATP 3-11.23/MCWP 3-37.7/NTTP 3-11.35/AFTTP 3-2.71. *Multi-Service Tactics, Techniques, and Procedures for Weapons of Mass Destruction Elimination Operations.* 1 November 2013.

ATP 3-11.32/MCWP 10-10E.8/NTTP 3-11.37/AFTTP 3-2.46. *Multi-Service Tactics, Techniques, and Procedures for Chemical, Biological, Radiological, and Nuclear Passive Defense.* 13 May 2016.

ATP 3-11.36/MCRP 10-10E.1/NTTP 3-11.34/AFTTP 3-2.70. *Multi-Service Tactics, Techniques, and Procedures for Chemical, Biological, Radiological, and Nuclear Planning.* 24 September 2018.

ATP 3-11.37/MCWP 3-37.4/NTTP 3-11.29/AFTTP 3-2.44. *Multi-Service Tactics, Techniques, and Procedures for Chemical, Biological, Radiological, and Nuclear Reconnaissance and Surveillance.* 25 March 2013.

ATP 3-11.41/MCRP 3-37.2C/NTTP 3-11.24/AFTTP 3-2.37. *Multi-Service Tactics, Techniques, and Procedures for Chemical, Biological, Radiological, and Nuclear Consequence Management Operations.* 30 July 2015.

ATP 3-28.1/MCWP 3-36.2/NTTP 3-57.2/AFTTP 3-2.67. *Multi-Service Tactics, Techniques, and Procedures for Defense Support of Civil Authorities (DSCA).* 25 September 2015.

TM 3-11.32/MCRP 10-10E.5/NTRP 3-11.25/AFTTP 3-2.56. *Multi-Service Tactics, Techniques, and Procedures for Chemical, Biological, Radiological, and Nuclear Warning and Reporting and Hazard Prediction Procedures.* 15 May 2017.

TM 3-11.91/MCRP 10-10E.4/NTRP 3-11.32/AFTTP 3-2.55. *Chemical, Biological, Radiological, and Nuclear Threats and Hazards*. 13 December 2017.

WEB SITES

Chemical Weapons Convention. <http://www.cwc.gov/>, accessed 17 October 2018.

PRESCRIBED FORMS

This section contains no entries.

REFERENCED FORMS

Unless otherwise indicated, DA forms are available on the Army Publishing Directorate (APD) Web site: <https://armypubs.army.mil/>.

DA Form 2028. *Recommended Changes to Publications and Blank Forms.*

RECOMMENDED READINGS

ADP 3-05. *Special Operations.* 29 January 2018.

ADP 3-07. *Stability.* 31 August 2012.

ADRP 5-0. *The Operations Process.* 17 May 2012.

ADRP 6-0. *Mission Command.* 17 May 2012.

Allied Tactical Publication 3.8.1, Volume III. *CBRN Defence Standards for Education, Training, and Evaluation.* 5 April 2011. Web site <https://nso.nato.int/protected/nsdd/APdetails.html?APNo=825&LA=EN>, accessed 15 May 2019.

ATP 1-05.01. *Religious Support and the Operations Process.* 21 July 2018.

ATP 3-90.4. *Combined Arms Mobility.* 8 March 2016.

ATP 4-02.7/MCRP 4-11.1F/NTTP 4.02.7/AFTTP 3-42.3. *Multi-Service Tactics, Techniques, and Procedures for Health Service Support in a Chemical, Biological, Radiological, and Nuclear Environment.* 15 March 2016.

ATP 4-02.83/MCRP 4-11.1B/NTRP 4-02.21/AFMAN 44-161(I). *Multiservice Tactics, Techniques, and Procedures for Treatment of Nuclear and Radiological Casualties.* 5 May 2014.

ATP 4-02.84/MCRP 4-11.1C/NTRP 4-02.23/AFMAN 44-156_IP. *Multiservice Tactics, Techniques, and Procedures for Treatment of Biological Warfare Agent Casualties.* 25 March 2013.

ATP 4-02.85/MCRP 4-11.1A/NTRP 4-02.22/AFTTP(I) 3-2.69. *Multi-Service Tactics, Techniques, and Procedures for Treatment of Chemical Warfare Agent Casualties and Conventional Military Chemical Injuries.* 2 August 2016.

ATP 5-19. *Risk Management.* 14 April 2014.

DODD 1300.22. *Mortuary Affairs Policy.* 30 October 2015.

DODD 5160.05E. *Roles and Responsibilities Associated with the Chemical and Biological Defense Program (CBDP).* 8 September 2017.

DODI O-2000.16, Volume 1. *DOD Antiterrorism (AT) Program Implementation: DOD AT Standards.* 17 November 2016.

FM 1-05. *Religious Support.* 21 January 2019.

FM 2-0. *Intelligence.* 6 July 2018.

FM 3-18. *Special Forces Operations.* 28 May 2014.

JP 1. *Doctrine for the Armed Forces of the United States.* 25 March 2013.

JP 1-04. *Legal Support to Military Operations.* 2 August 2016.

National Fire Protection Association 472. *Standard for Competence of Responders to Hazardous Materials/Weapons of Mass Destruction Incidents.* Current Edition 2018. <https://www.nfpa.org/codes-and-standards/all-codes-and-standards/list-of-codes-and-standards/detail?code=472>, accessed 13 March 2018.

Section 120, Part 1910, Title 29, Code of Federation Regulations. *Occupational Safety and Health Administration, Department of Labor.* <https://www.ecfr.gov/cgi-bin/text-idx?SID=9440390c6578b6d8303e57b6f4a7278d&mc=true&tpl=/ecfrbrowse/Title29/29tab_02.tpl>, accessed 17 October 2018.

STANAG 2515. *Collective Protection in a Chemical, Biological, Radiological, and Nuclear Environment (COLPRO), Edition 2. (ATP-70, Edition A, Version 1).*1 February 2014. Web site <https://nso.nato.int/protected/nsdd/stanagdetails.html?idCover=8356&LA=EN>, accessed 17 October 2018.

STANAG 2520, *CBRN Defence Standards for Education, Training, and Evaluation, Edition 1* (ATP-3.8.1, Volume III). 27 November 2011. Web site <https://nso.nato.int/protected/nsdd/stanagdetails.html?idCover=7208>, accessed 17 October 2018.

STANAG 2521. (ATP 3.8.1, Volume 1). *CBRN Defence on Operations. Edition 1.* 31 January 2009. Web site <https://nso.nato.int/protected/nsdd/stanagdetails.html?idCover=7209&LA=EN>, accessed 17 October 2018.

STANAG 3497 (AAMedP-1.8, Edition A, Version 1). *Aeromedical Training of Aircrew in Aircrew CBRN Equipment and Procedures. Edition 4.*9 January 2018. Web site <https://nso.nato.int/protected/nsdd/stanagdetails.html?idCover=9048&LA=EN>, accessed 17 October 2018.

TM 3-11.42/MCWP 3-38.1/NTTP 3-11.36/AFTTP 3-2.83. *Multi-Service Tactics, Techniques, and Procedures for Installation Emergency Management.* 23 June 2014.

Index

Entries are by paragraph number.

Made in the USA
Coppell, TX
11 November 2020

41139572R00070